The Theories
Of
Lenard Metzger

By
Lenard Metzger

Elemental Publishing
Rochester, New York

Other books by Lenard Metzger

COMMON SENSE COSMOLOGY

BEYOND EINSTEIN

THE THEORIES OF
LENARD METZGER

Elemental Publishing
80 Westerloe Avenue
Rochester, NY 14620
(585) 473-9303

Library of Congress
Control Number: 2010930673

ID: 8816601
Lulu.com

ISBN: 978-0-557-51450-2

PREFACE

This book is a continuation of my previous book, "Beyond Einstein", which was a continuation of my first book, "Common Sense Cosmology". I hope that the readers of this book will have read them. In case they haven't, I have provided most of their content as addenda at the end of this book. Herein, I have concentrated on the most significant chapters and added more details and discussion to them. I will no longer devote much time to justify my disavowal of the two theories of relativity. I now realize that I tried to cover too much in these earlier books. I included sections on cosmology, astrophysics and even nuclear physics, plus paintings, poetry and fantasy.

It was my intention that "Beyond Einstein" would be understandable and persuasive. It has gone through all the procedures of registration, publicity and distribution, without a bit of interest being shown by anyone. Copies of the book have been sent to the heads of over twenty departments of astrophysics of major universities throughout the United States, without a single response. I no longer believe that this augmented version will be received with any greater enthusiasm. Since I am over 83 years of age, I just want to finish this book and be done with it. If at some later time it is discovered, after I am gone, so be it.

THE THEORIES OF LENARD METZGER

TABLE OF CONTENTS

CHAPTER	PAGE
Introduction	6
Black Holes	8
Big Bang	12
Universe	15
Dark Energy	19
Gravity	20
Weight	22
Neutrons	24
Photons	26
Electrons	28
Life	35
Proof	36
ADDENDUM A	41
"Beyond Einstein"	
ADDENDUM B	153
"Common Sense Cosmology"	
Excerpts	

INTRODUCTION

I believe my most important ideas are in astrophysics. They began with my desire to reconcile the dichotomy between the electro-magnetic and the quantum theories of light. My assumption about the nature of dark matter, to explain the speed of light, was key. This concept of dark matter led me to a new explanation for gravity. This also led to explanations for black holes and the big bang.

Dark matter is believed by science to make up most of the mass of the universe. I proposed that dark matter consists of neutrons traveling at or near the speed of light. I suggested that they actually propel photons to that speed.

In another important theory, I describe how the effect of gravity is caused by the weak interaction of these particles with matter. In my previous books I had called these speed-of-light neutrons, "gravitons" to distinguish them from the familiar, low velocity neutrons usually detected. However, the term "gravitons" has a different connotation in physics so I have taken advantage of my literary license and given them a new name. With no false modesty, I will name the speed-of-light neutrons "metz " after myself. Hereafter, the speed of light will be known as the speed of metz.

In my theory of how black holes are formed I proposed that in addition to neutrons, as in neutron stars, other forms of neutral particles are created in the center of larger black bodies.

The collapse of a large star into a black hole will stop at a finite volume with particles, made of charm and strange quarks, forming in the center. I named these particles, "charstrons". Each would have the mass equivalent to about a hundred neutrons.

I also proposed that a third kind of neutral particles would be made of top and bottom quarks and would have the mass-energy of about a hundred charstrons. They would form in the center of much larger black bodies, such as the one at the center of our galaxy. I named these particles, "tobotrons".

BLACK HOLES

You can see the full chapter on black holes starting at page 125. It seemed best to begin, in describing the evolution of black holes, by going one step backwards to neutron stars (which I renamed "neutroids" since they are not stars). My library research indicated that "exotic particles" were thought to occur at the center of neutroids. This triggered my idea that the exotic particles would be another form of neutral particle.

The idea of a solid body made almost entirely of neutrons, all in contact with one another, made me wonder if any of the metz impacting on its surface could make it through to exit the opposite side. This would seem to be necessary, for any photon impinging on the surface, to be reflected. A metz would propel it away and make the body visible. I do not think that metz can penetrate a body of solid neutrons. A possible explanation occurred to me, as suggested by the following figure.

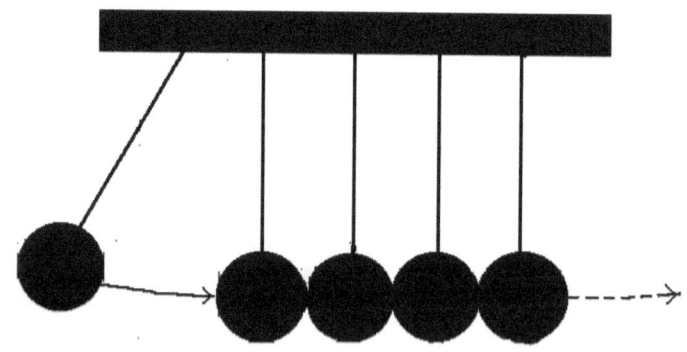

Transfer of Momentum

Figure A

From the above diagram, of a common demonstration object, it seems possible that the momentum of a metz, impacting on a surface neutron, could be transmitted through the neutroid. Thus causing a different neutron to be expelled from some other point on the surface. It would then become a metz capable of propelling a photon away from the surface. This may only occur for a small percentage of the impacts. Most of the metz would dissipate their energy as heat, and in applying pressure at the center. They would also add to the mass of the neutroid.

In the following figure the possible evolution of black bodies is described. They range from the black body, from a single large star, all the way to a single body containing all the mass of the entire visible universe. These different bodies in this diagram are only suggestive of possible sizes and proportions.

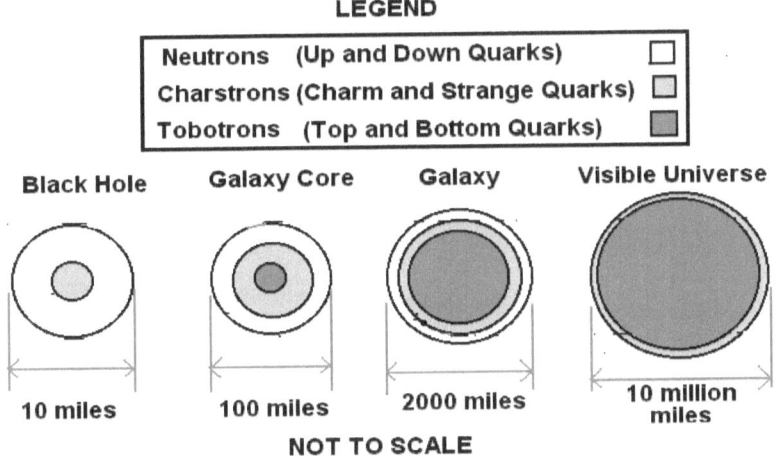

BLACK HOLE EVOLUTION

LEGEND

Neutrons (Up and Down Quarks)
Charstrons (Charm and Strange Quarks)
Tobotrons (Top and Bottom Quarks)

Black Hole   Galaxy Core   Galaxy   Visible Universe

10 miles   100 miles   2000 miles   10 million miles

NOT TO SCALE

Figure B

The black hole core is shown as charstrons. The size of the core would be such as to be able to sustain the neutron mass above it, without further collapse. Similarly, the galaxy core black hole is suggested to have a core of tobotrons, surrounded by a shell of charstrons and an outer shell of neutrons. As I indicated, the relative sizes of all the layers are to be determined.

A possible explanation of why these bodies are black is that light cannot reflect from them because no metz leave them to propel the photons away. The following figure suggests why this may be so.

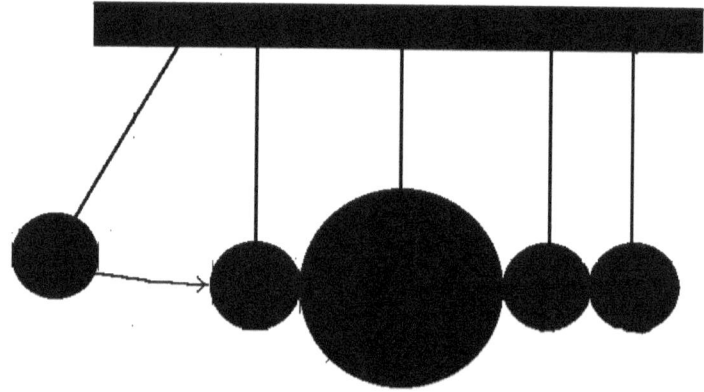

Transfer of Momentum?

Figure C

It seems unlikely that the impact of the small ball at the left would transfer enough momentum through the large ball in the center to move the small ball on the right. Similarly, the impact of metz on the outside of the above black hole would transfer their momentum into the charstron core and not through it.

The momentum of a metz impact could not get back out to the surface of the body. Photons would not be propelled away and the body would be invisible.

BIG BANG

As in the chapter starting on page 133, there are several possible explanations for what triggered the big bang. An impact with another massive black body could have cracked it open. It could have become too massive and collapsed. However, I favor the concept that when the flux of metz, impacting on its surface, fell below a certain limit, the internal pressure opened up the outer shells and the explosion took place.

The following figure shows how the big bang might have looked, in a scaled down fashion. The idea was inspired by my recollection of a particular finale of a fireworks display. A single bomb was fired upward. It burst into a sphere of bomblets. Each of the bomblets burst into a large number of smaller bomblets. The entire sky was filled with the interlaced tracks.

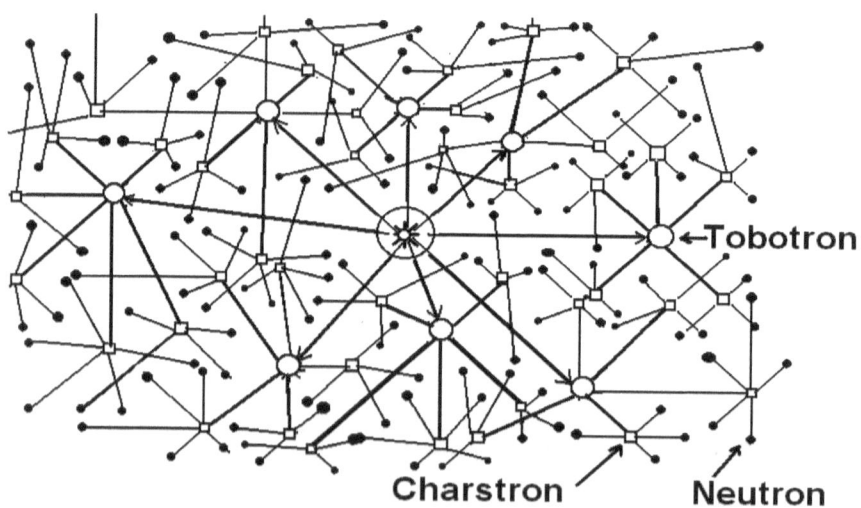

BIG BANG

Figure D

In the figure above, only four or five particles are shown leaving each explosion point, instead of the proposed 100. It can be seen that many of the final particles (neutrons) would be moving back towards the center of the big bang. I assumed that it was there that most of the neutron-to-neutron collisions occurred and produced the protons and electrons that would become the visible universe.

The figure below is from page 133 and shows a neuron-to-neutron impact. Three of the down quarks, each with a charge of one third, would combine into an electron, with a charge of unity. The two up quarks and the remaining down quark would combine into a proton.

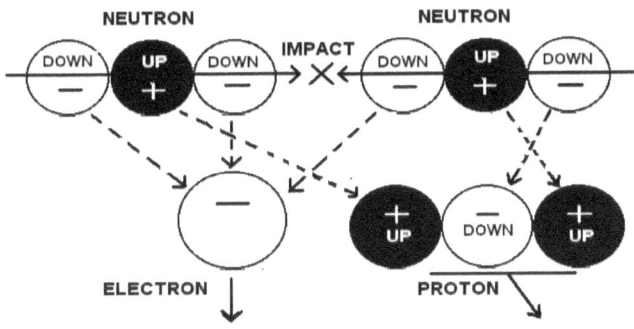

Figure E

The surplus mass of the down quarks would produce the energy to propel the resulting particles away. This would result in the center of the resulting universe being sparsely occupied.

The relatively thin outer shells of neutrons and charstrons of the universe-sized black hole (Univoid?) would have been propelled outward at faster than today's velocity of metz. These would become the tenuous outer reaches of the present universe.

Most of the mass of the universe (neutrons and protons) would have combinations of radial and tangential velocities and would produce the great majority of the distribution of galaxies, stars and dark matter, occupying the middle region of the present universe.

In the previous book I raised the question as to whether there would be a sufficient number of metz to account for gravity and the speed of light, throughout the universe. To answer this question requires an estimate of the size of the visible universe and the number of metz in it.

UNIVERSE

The following figure uses a method of showing my concept of the universe.

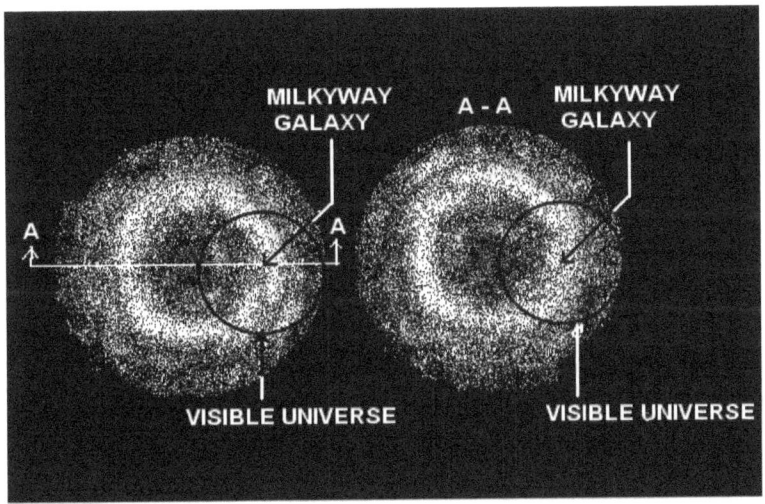

Figure F

The circle displayed on the left is a cross sectional view of the universe, in the plane of the paper. The section A-A at the right is a cross sectional view at right angles to the plane of the paper. Assuming that the age of the universe is 13 billion years, the diameter (roughly from A to A) would be about 26 billion light (metz) years. As I have previously explained, I reject the relativistic concept of expanding space.

As indicated by the smaller black circles, the visible universe would extend from the center of the universe to its outer reaches. This circle has a radius of about 6.5 billion metz years. The Milky Way galaxy is assumed to be near the center of this volume.

The density of matter along the radial direction is peaked near the center of the visible universe with a some-what- normal distribution. The distribution is more uniform along the circumference at a constant radius.

The Milky Way galaxy was assumed to contain 100 billion solar masses. The visible universe black body could have the mass of 100 billion galaxies. Therefore, the visible universe black body can be estimated as having 10 to the power of 22 (10^{22}) solar masses, at a minimum.

I assumed that what was to become dark matter increased the mass of the visible universe black body by a factor of eight. This doubled the diameter of the body to 20 million miles. The visible universe accounted for one eighth of this and the rest became dark matter. I assumed the dark matter, in the visible universe, approximated 10^{23} solar masses.

Using the mass of the sun (one solar mass) and the mass of a neutron, I calculated that a minimum of 10^{80} neutrons (metz) constitute the dark matter in the visible universe.

Referring to Figure F, The diameter of the full universe was assumed to be twice that of the visible universe. Therefore, the full universe would have a volume eight times that of the visible universe.

The original diameter of the full universe size black body would be about 40 million miles. At the big bang it would expand, almost instantaneously, to about a billion miles in diameter.

In calculating the volume of the visible universe, an average radius of about 5 billion metz years was assumed. Using the equation for volume of a sphere, 4/3 (pi) R^3, the visible universe comes out a sphere with a volume of about 5 X 10^29 cubic metz years.

A cubic metz year is an awkward unit to use for this purpose. Instead I repeated the calculation in a unit I believe is more useful, a cubic metz second. The number of seconds in a year is about 3 X 10^7. Therefore, for the visible universe, I used a radius of 15 X 10^16 metz seconds.

This gives a volume of about 5 X 10^51 cubic metz seconds. A metz second is 3 X 10^8 meters, which is 3 X 10^14 microns. Each face of a cubic metz second is about 10^17 square meters and also 10^29 square microns. This is shown in the figure below.

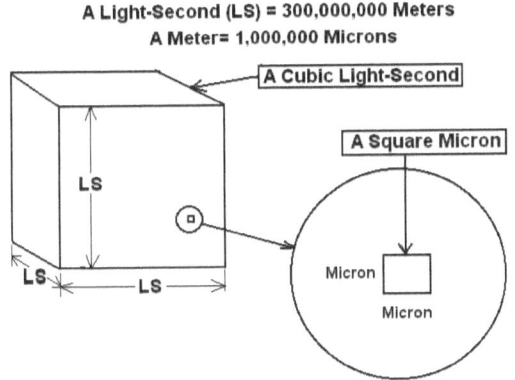

Figure G

Metz entering the above cubic metz second, normal to the face will pass through all 3 X 10^14 cubic microns in the column, in one second.

It would take a metz entering each of the square microns on the face, 10^{29} of them, to pass through every cubic micron in this cubic metz second, every second.

For all the cubic metz seconds, throughout the entire volume of the universe, to be visited by metz every second, would take about 5×10^{80} metz. This requirement is a factor of 5 more than the number of metz computed to be in dark matter, but this is within the margin of error for this calculation.

The available number of metz would appear to be enough if we set the dimensions of the elemental cube to about 2 microns. It is obvious that all metz will not be oriented perpendicular to every cubic surface at the same instant, as assumed above. The random orientations and timing of the passage through space of every metz, will make their visit to each atom highly variable.

It remains to be determined how close a metz has to come to an atom to trigger the emission of a photon or how close to a photon to accelerate it to its characteristic velocity. It also depends on how close a metz has to come to an atom of a body to impart a bit of momentum to the atom, through electro magnetic interaction, causing deflection of the atom and the metz, thus producing gravitational effects.

DARK ENERGY

I consider dark energy to be the kinetic energy possessed by all of the dark matter particles traveling at their characteristic velocity. This velocity averages the present speed of light. Reference material, that I have found, calls these particles, "wimps". This stands for "weakly interacting massive particles". I prefer the name metz instead of wimps.

The amount of visible matter is stated to be less than 4% of the total mass of the universe. The dark matter is estimated to be about 23% and the dark energy is the remaining 73%.

If one insists on using the equations of relativity, for the hypothetical rest mass of the dark matter to be increased to a relativistic mass that is four times as great, their velocity would have to be within about 3% of the speed of "light".

This was calculated using the following equation.
$M_r = M_o / [\{1-(v/c)^2\}^{1/2}] = 4M_o$
Or $1/16 = 1-(v/c)^2$
And $v/c = (15/16)^{1/2}$
Therefore $v = .968c$

A classical explanation is to consider the kinetic energy of a metz as:
$\frac{1}{2}(M_n) \times (V_c)^2 = \frac{1}{2} (1.67 \times 10^{-27}) \times (3 \times 10^8)^2$
This equals 7.5×10^{-11} joules, as the kinetic energy of a metz. The total dark energy of the visible universe would be 7.5×10^{69} joules.

GRAVITY

METZ - ATOM INTERACTION

The assumption that more force will be transferred by metz to matter in proportion to the density of the matter needs examination. (See page 70.)

For a given volume of matter, it is obvious that the closer together the atoms are to each other, the greater will be the weight of the matter, with a particular atomic weight. This is because the probability that metz will interact with atoms goes up as the space between the atoms is reduced.

However, if all things are equal, except the atomic weight of the atoms, the weight of a volume of matter goes up in proportion to the weight of the atoms.

The diagram below shows the electro-magnetic interaction between metz and atoms with different atomic weights. There should be a change in the amount of deflection caused to metz when interacting with atoms with different atomic weights. Different amounts of momentum will be transferred, in the direction of the metz travel, to the atoms of different weights.

The following figure shows how this may occur. A metz is not deflected appreciably in moving a light-weight atom out of its way. Interacting with a heavy atom causes the metz to be deflected, in imparting momentum to the atom.

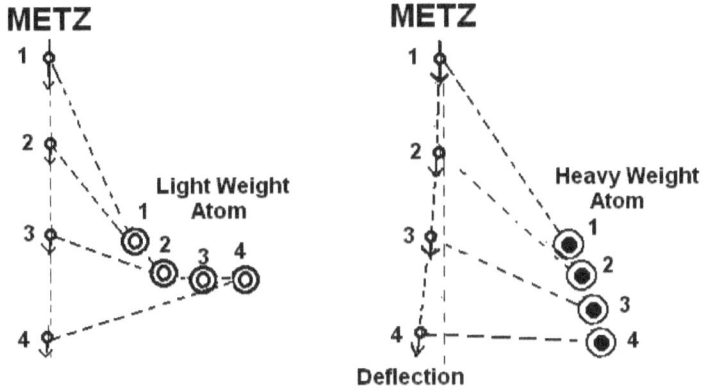

INTERACTIONS AT DIFFERENT TIMES

Figure H

Metz passing on both sides of these atoms would cause mirror image results. The lightweight atom would be moved from side to side, with very little vertical motion. The heavy atom would gain additional vertical momentum with each interaction.

WEIGHT
(See page 73.)

The following figure should be considered in calculating the weight of objects in the vicinity of the Earth.

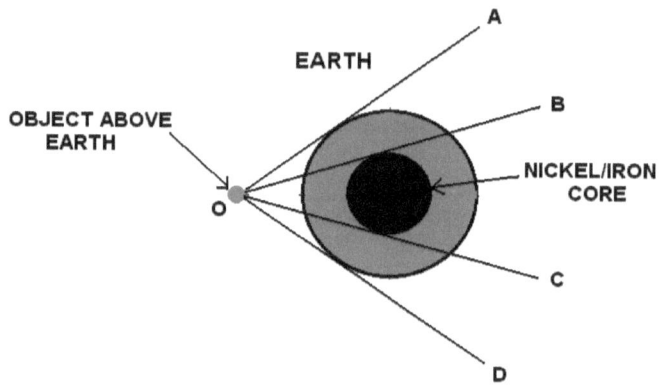

Figure I

I think that the atoms, in the dense nickel-iron core of the Earth, could have many close encounters with a metz. This would result in many of the metz being deflected and transferring momentum to the core. This could account for the high pressure on the core and its high temperature.

As a first approximation, the core could be considered as blocking most of the metz coming through the solid angle, subtended by the core, at the small object.

Therefore, most of the momentum from the metz propelling the small object towards the Earth would be coming from the direction opposite to the Earth, through that same solid angle.

It is obvious that any metz whose path would have intercepted the small object, if the metz were not deflected, would miss the object, if it were deflected. This will occur 100% of the time.

The question remaining is; how many of the metz that would have missed the small object, if the metz were not deflected, would intercept the small object if the metz were deflected? In the next figure I address that question.

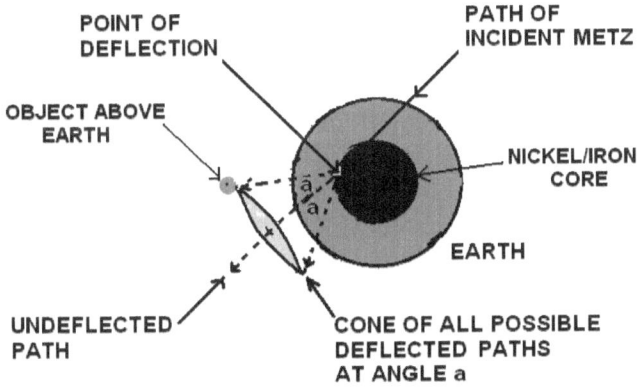

Figure J

A metz is shown on a path that if it were to continue would have it missing the object. A deflection point is shown that for a specific angle a, and only one of the many possible deflections, the metz will intercept the object. Obviously the probability of this occurring is extremely low.

NEUTRONS

These theories all revolve around the neutron. My concept of the configuration of the quarks in a neutron is based on the fact that the quarks have electric charge and electric fields. Due to their spinning charge, they will have magnetic fields. These fields should hold the quarks together. The following figure shows how the magnetic fields tend to do this.

NEUTRON

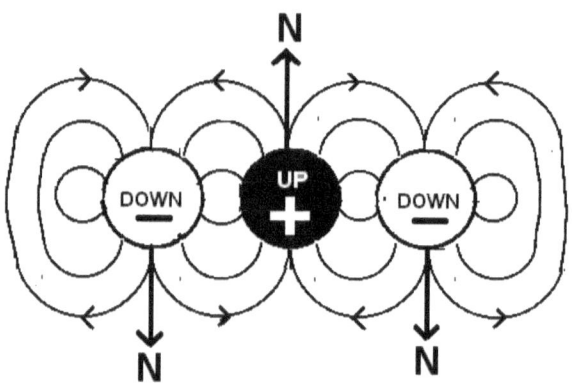

Magnetic Fields

Figure K

One could lay three bar magnets side by side with the polarities arranged as those above and they will move together and hold each other firmly.

In the figure below, I present my recollection of how the electric force fields surround oppositely charged bodies. These alternate charges will also tend to hold the quarks together.

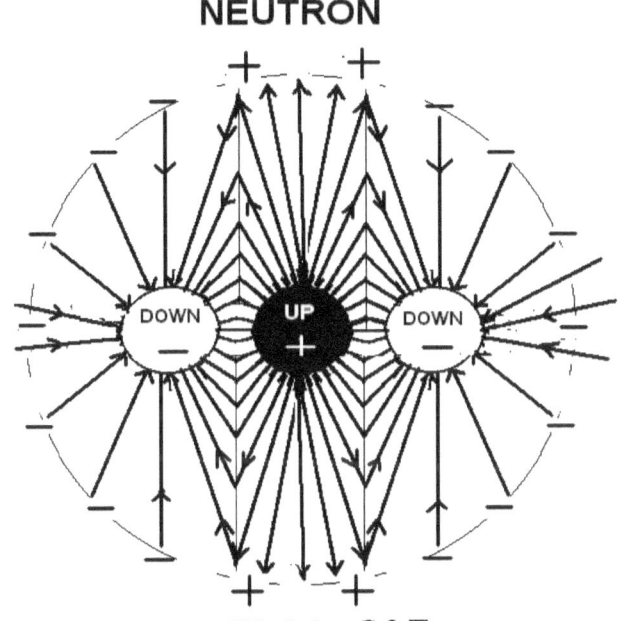

NEUTRON

Electric Fields Of Force

Figure L

Because the positively charged quark is bracketed by two negatively charged quarks, I assume that the surrounding, superimposed electric fields of the neutron would be more negative than positive around the circumference. As a neutron is racing past an object, the spinning fields would effect the object more by the longer duration, negative, force fields than by the shorter duration, positive ones. If the spin is not considered significant, the probability is greater that a negative field will interact with an object, in passing.

PHOTONS

The figure below is from page 81 in the chapter on light.

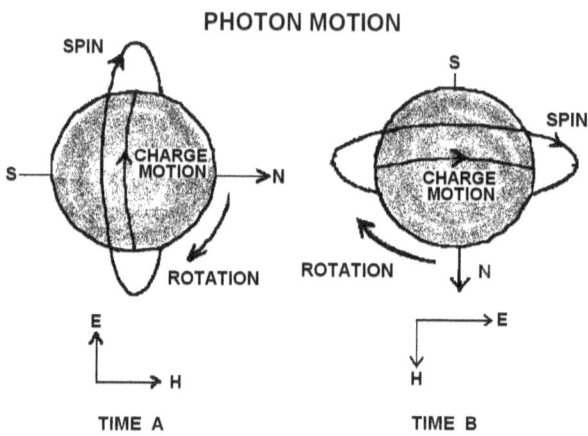

Figure M

This concept of photons assumes that they are tiny charged bits of the electrons emitting them. They also have retained some of the electron's spin. This provides their magnetic fields. Their rotation will provide their moving electric fields.

In the chapter on hydrogen, the figure on page 97 shows how the interaction of an electron with the nucleus can cause a bulge to build up on the electron. A passing metz may then trigger the release of a photon.

The following figure is from the chapter on gravitons, now called metz. See page 137. It shows how a metz in passing a photon, either head on, or from the rear, will deflect the photon.

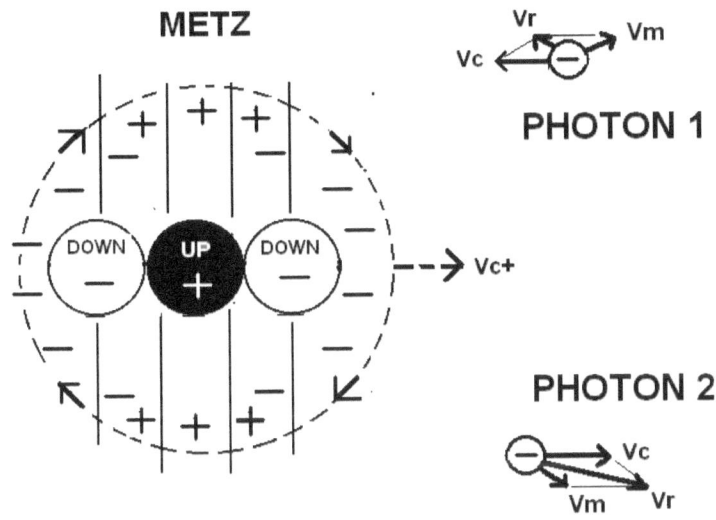

Figure N

The photon number 1 is moving towards the metz and its reaction to the metz's fields is to slow down and move away. The photon number 2 is moving in the same direction as the metz and it will speed up and move away from the metz.

ELECTRONS

In the development of my idea of the emission of light photons, I assumed a small bulge would build up on the surface of an electron and fly off as a photon. In thinking about this I began to wonder if the electron consists only of "charge" or if the "charge" is a layer or coating on some core body of the electron. I have derived an equation that indicates that a photon of visible light, with a wavelength less than a micron, has a mass slightly larger than a millionth of the mass of an electron. However, gamma rays have masses many orders of magnitude greater than a visible photon. What happens to an electron as it emits a gamma ray?

Gamma rays are said to have wavelengths of less than 10^{-5} microns. This means that gamma rays are at least 10^5 times more massive than a visible photon. Therefore, a gamma ray would have, at least, a mass of about a tenth of an electron's. It would seem that the total mass of an electron would determine the shortest wavelength of a gamma ray. By these calculations, the gamma ray wavelengths should cover a range of ten to one.

The electron is arbitrarily defined as having a negative charge, of a unit value. A proton is defined as having a positive charge, equal but opposite to the negative charge of the electron. The proton has a mass a factor of over a thousand times greater than that of an electron. This would have raised the question of the density of positive "charge" or if a more massive core proton particle was coated with a similar amount of charge, as is on the electron.

The quark theory of nuclear particles complicates this discussion. It states that protons and also neutrons are made up of two different kinds of charged particle. One particle, called the "up" quark, if said to have a positive 2/3 charge. The other particle, called the "down" quark, is said to have a negative 1/3 charge.

A proton is made up of two "up" and one "down" quark. A neutron is made up of two "down" and one "up" quark.

The mass of the proton (and the neutron) is a complex function of the individual masses of the quarks and the energy equivalent mass of their interactions. In any case the mass of the negative quark appears to be much greater than the mass of the electron, even with a smaller charge than the electron.

What is at the core of these particles and what is the difference in consistency between positive and negative charge? But that is only part of the problem. What is the electric field that emanates from the charge? What is the difference between positive and negative fields that make them attract each other? What makes two similar fields, positive with positive and negative with negative, repel each other?

One thought that I have had is that a charge, coating the core of a particle, could extend out into the volume of space surrounding the particle, with the density of the charge decreasing with distance at the rate computed by the field equations. There still is the question of why the like charges push each other away and the opposite charges pull together and intermingle.

I believe there is a physical reality underlying these questions. We may never be able to "put our fingers on it" but there is something there besides the mathematical relationships that we use to describe it.

The following discussion is highly speculative. I do not have much confidence in its validity. However, I believe that this is an area of science that needs study.

If I were designing an electron it might look something like what I show in the following figure.

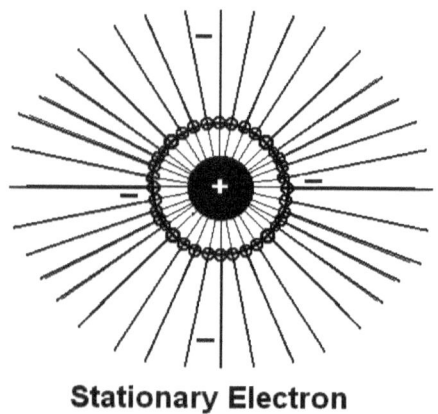

Stationary Electron

Figure O

The dark inner sphere could be a neutral or of a positive body, as shown, to hold the outer shell of small spheres, each with a long radial tail. These would be the elemental units of the negative charge and also constitute the analog of the electric field. These long tails would extend much farther out than is shown. The small spheres would act as signal processors, analogous to the nuclei of nerve cells. The long tails would act as sensors.

When two similar objects approached each other, the sensors of each would make contact and produce a repelling reaction. If the contact were made with an object of the opposite polarity, the reaction would be to fasten onto each other's sensors and retract. This would draw the objects closer together. As they approach each other the number of sensors making contact would increase and the rate of closure would increase. This is the analog of the electric field strength varying inversely with distance.

If the tips of the sensors were to rotate, one kind of charge clockwise and the other kind in the opposite direction, the like kinds of sensors would repel each other. When the opposite kinds contacted they would wind around each other and draw closer together.

I have described the concept of photons being small bits of the electron emitting them. I assumed that the electron had a fluidic or granular nature. By the above picture of an electron, it could be considered granular. This would imply that various photons of differing energy, hence differing masses, would contain differing numbers of these elemental cells. A photon with a single element, as in the figure below, would be that with the lowest mass possible.

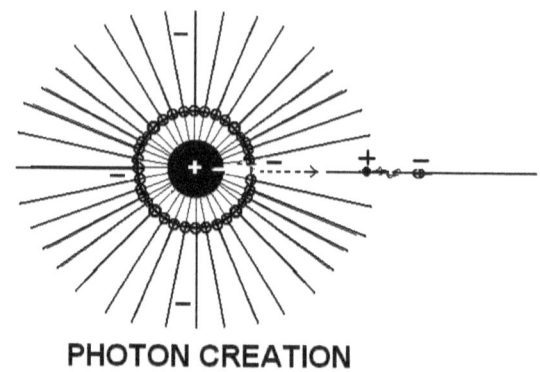

PHOTON CREATION

Figure P

A minimum energy photon, as above, would probably be one in the infrared region with a wavelength of from ten to a hundred microns. A visible photon, with a wavelength of half a micron, would be a factor of about 10^2 times more massive.

Since a visible photon was assumed to have a factor of 4×10^{-6} less mass than an electron, therefore an electron can be assumed to have over 10^8 of these sensing elements.

It remains to be determined if this model of an electron is at all feasible. The question also remains whether the core would provide the mass of the electron or if the shell of small cells would provide most of the electron's mass.

The other characteristic of an electron that is much more difficult to explain, is the "magnetic field" generated when the charge moves. Why does a moving charge produce a magnetic field? What is a magnetic field?

Consider something moving a charged particle by pushing on the periphery of the field. What is the motion of the core of the particle? Does its inertia warp the charge? Does the charge curl into circular vortices?

The motion of the charge can be due to the spinning of the electron. In the following figure I show how the spin of the electron might produce a curvature to the long tailed sensors.

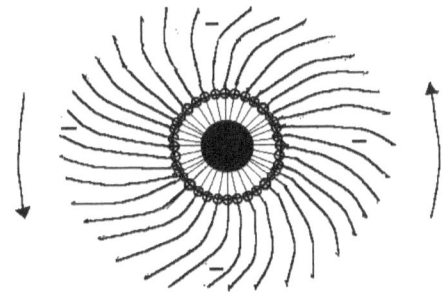

Rotating Electron

Figure Q

I haven't a good idea how this would produce a magnetic field. The magnetic lines of flux would have to appear somewhere along these curved sensors, passing vertically through the plane of the paper. Perhaps the ends of the sensors would coil.

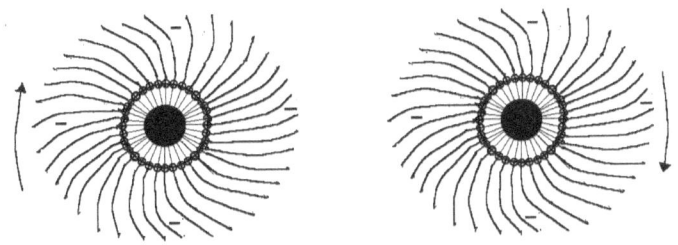

TWO ELECTRONS SPINNING CLOCKWISE

Figure R

In the above figure, two electrons are shown spinning in the same direction. They could be considered as having the same magnetic polarity. Their curved sensors would impinge onto each other and propel them apart.

In the figure below, the two electrons are shown spinning in opposite directions. This could be considered as if the electrons had oppositely polarized magnetic fields. Their curved sensors would move into each other and pull them closer together. I envisioned this as being analogous to a two handed shuffle of a deck of cards.

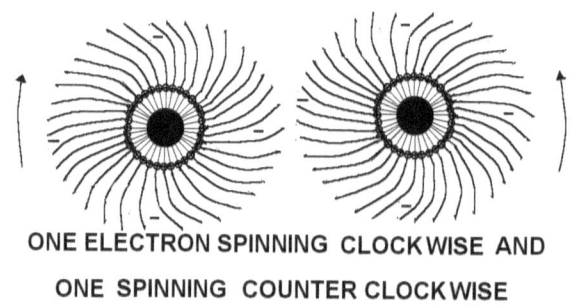

ONE ELECTRON SPINNING CLOCKWISE AND
ONE SPINNING COUNTER CLOCKWISE

Figure S

I think that these concepts are interesting but are probably not realistic. However, some thought should be given to finding the actual explanation of these phenomena and not be satisfied with just a math model of them.

LIFE

The question may well be asked, as to how living tissue can tolerate being subjected to speed of light particles passing through every cubic micron, continually. It has been established that neutrinos from the sun do just that. Since they are non-ionizing particles, neutrinos have not been thought to produce health hazards.

Photons, from ultra violet, to X-ray to gamma ray are all ionizing radiation and above a certain concentration do cause tissue damage.

Since living objects, from microbes to humans, have all evolved under the influence of gravity, if my theory of gravity being caused by metz is true, all Earthly life has developed ways to tolerate this exposure. It may even be that cells make use of the passing metz to cause desired functions, such as energy production.

PROOF

Since I wrote the "highly speculative" chapter on faster-than-light travel, starting on page 143, I have been increasingly doubtful of its possibility. This is mostly because I cannot picture sufficient numbers of head on collisions of metz occurring to produce sufficient protons and electrons to give adequate thrust to propel a space- ship.

If a stationary, experimental setup could be developed that produced these collisions, under controlled conditions, it would go a long way to verify my basic assumptions about metz and dark matter. I anticipate that the collisions would occur only occasionally. I assume that the metz would usually tend to repel each other, even when they are on an exact, head on path. It seems that some thing might be done to orient them, so they would attract each other. I show two possibilities below

.

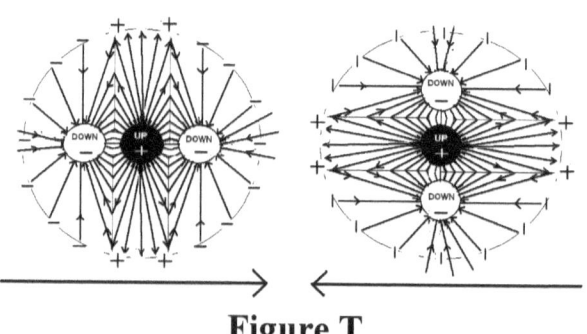

Figure T

In the above configuration the maximum positive electric field of one metz would face the negative electric field of the other, just before impact. Another possibility is shown in the following figure.

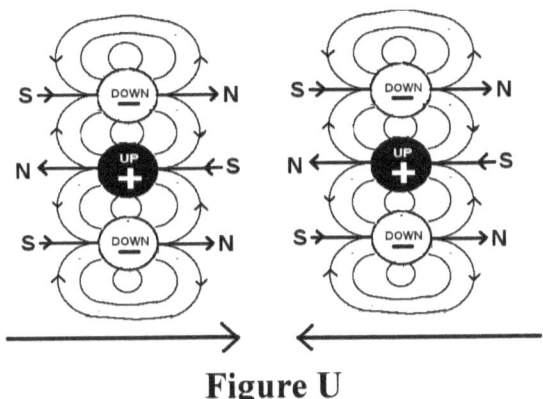

Figure U

In Figure U, the magnetic poles of the quarks in one metz, face the opposite polarity magnetic poles of the quarks in the other metz. This should produce an attractive force, between them, just before impact. The questions remain as to how to produce these final configurations and if this collision will produce a proton-electron pair. The following is an experimental setup to try.

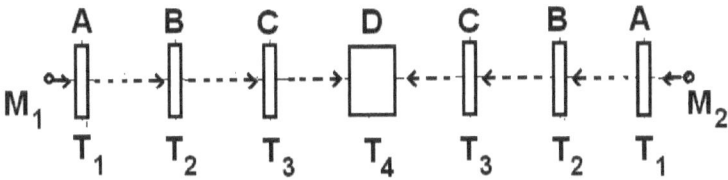

METZ EXPERIMENT

Figure V

In the operation of the set up of Figure V, two metz, M1 and M2, will enter the "A " elements, at the opposite ends of the apparatus. This must occur at approximately the same instant, during the activation of the A elements, at time T1.

The B elements will be activated at a time T2, after a time delay that will be determined by the distance from A to B and the time taken by the metz to travel this distance. As an example, if the distance between them is three meters the time delay would be 10^{-8} seconds (10 nanoseconds).

The activation of the C elements will occur at T3 after the proper time delay. The element D will be a particle detector for the expected products of the metz collision. It will be activated at time T4.

The three elements on each side of the detector are only for illustration purposes. The number required might be more, but probably not less than three. These units will be designed to cause the metz to travel along the axis of the setup and arrive at the collision point with the desired orientations for maximum collision probability.

I assume that the function of these units will involve pulsed magnetic and electric fields. The pulse widths would probably be about half that of the time delay durations. The particle detector should be long enough to allow some variation in the location of the collisions. There will be some variation in the times the metz enter the apparatus, during the pulse at time T1.

The repetition rate of this procedure will also depend on the reset time of the detector and the recharging of the electronics providing the pulses. It is likely that the detection of a collision will occur only rarely. Although the number of metz coming from all directions is extremely high, the number that satisfy the conditions for this experiment will be very small.

The apertures of the devices will be limited to a size that will allow adequate pulsed field strengths to occur during the short duration pulses. A couple of possible device configuration to produce various electric fields is shown below.

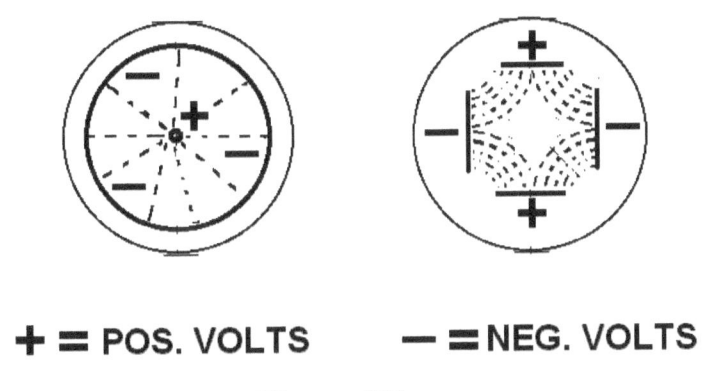

+ = POS. VOLTS   — = NEG. VOLTS

Figure W

In the figure above, the diagram on the left shows a radial electric field. The small anode in the center has a positive voltage. The circular cathode surrounding it has a negative voltage. Of course these polarities could be reversed.

The diagram on the right shows a more complex electric field. Two pairs of plates are shown but there could be more than this number. Several of this type of device, placed in sequence, might control the spin of a metz.

Simple current loops could be used to generate magnetic fields concentric with the device axis. These fields might polarize the metz to the proper orientation and move them to the axis.

ADDENDUM A.

BEYOND EINSTEIN

By

Lenard Metzger

Lulu

Other book by Lenard Metzger

COMMON SENSE COSMOLOGY

ID: 332616
lulu.com

BEYOND EINSTEIN

Library of Congress
Control Number: 2008903425

Published by
Lulu Enterprises
Morrisville, NC 27560
United States of America

ID: 1681945
lulu.com

ISBN 978-1-4357-1461-8

PREFACE

How could people have believed that the world was flat and that if you sailed a ship to the edge your ship would fall off? Why didn't they realize that all the water in the seas would pour off the edge and the ship would have nothing to sail on?

After it was generally accepted that the world was round, there was this idea that the Sun and the stars all revolved around the Earth. And even when it was finally understood that the Earth revolved around the Sun, the belief was that the stars were on the surface of a crystalline sphere that rotated around us. How could people have believed in crystalline spheres with nothing beyond them?

I have tried to visualize this, but every time I reached a mental barrier in space I continued outward from the other side of the barrier. I have concluded that there is no end to space.

A theory of black holes shows mathematically that all the mass of a large star can shrink down to zero size and still have an extremely large mass. Similarly, a theory of the beginning of the universe assumes that all of the mass of the eventual universe started out with zero size. How can people believe this?

The above is a condensed version of some of the preface of my previous book, "Common Sense Cosmology". It was a personalized presentation. It consisted of factual chapters, but also included fanciful sections. I had hoped that it would be understandable to non-technical readers. This was not accomplished very well.

In this book, I have eliminated most of the superfluous sections. I am concentrating here on my factual, if somewhat speculative concepts. I have added chapters with the ideas that were worked out during the interim period. I intend to use a more rigorously technical presentation.

It has become obvious that my ideas are in conflict with some of the theories of Albert Einstein. I will describe where I think Einstein erred. Not being restricted by his theories allowed me to develop new solutions to several basic problems. I understand that this is presumptuous, but I have no other choice but to have my say.

By the definitions and equations that Einstein used in his theories of relativity he prevented himself from solving the problem he spent most of his life working on. That was to unify the theories of electromagnetism and gravity.

I will present, in this book, a theory that gravity is the result of a repulsive force imparted by particles having a velocity greater than that of light. These particles also determine the speed of light. I call these particles gravitons, and later describe what I believe them to be. This is not the conventional definition of gravitons.

BEYOND EINSTEIN

CONTENTS   PAGE

Introduction   47
Space-Time   49
Starlight   51
Universe   55
Multi-Verse   61
Gravity   67
Weight   73
Light   77
Radio   83
Hydrogen   91
Stars   103
Black Holes   125
Big Bang   133
Gravitons   135
Motion   141
Faster-Than-Light   143
Conclusions   149

INTRODUCTION

I consider cosmology to be the study of everything, including our universe and its place in the cosmos. We will consider our universe, as it appears to exist at this time, and then look at where it came from and where it is going. To be able to do this we will have to establish some fundamental assumptions.

I believe that it is safe to assume that our universe began with something that was somewhat like what is called the "Big Bang". An expanding universe was postulated based on the red shift of various stars and galaxies, as a function of their distance from the Earth. Hubble and others quantified this relationship. The age of the universe was estimated at between 10 and 20 billion years. This was arrived at by projecting the estimated rate of expansion (the Hubble Coefficient) backward in time to where all the observed bodies seemed to have coincided.

The present theory of the earliest time of the universe is that the temperature was so high initially, that only gravitational energy existed. As the expansion continued and the temperature dropped, successive energy régimes occurred; strong force, weak force and electro-magnetic force. Then subatomic particles appeared, followed by the first atoms, hydrogen, deuterium and helium. Over vast periods of time, volumes of the hydrogen gas converged, becoming stars that combined into galaxies.

The above concepts will be the starting point for my new ideas. The approach that will be taken is to use the least amount of mathematics possible and use simple logic to describe my vision of how the cosmos works.

I will assume most generally accepted astronomical and physical facts as givens. Conclusions that have been reached as the result of manipulations of derived equations will be questioned and subject to re-interpretation. There are a number of prevalent theories that are not so easily understood or visualized. In some cases they are not even logical. In the following sections I will address these cases in detail and offer alternative explanations.

What I question is using a mathematical model of a phenomenon that requires adjusting the coefficients of the equations of the model, to give the desired answers. Even worse is extending the predictions of the model well beyond its normal range of data. Such a model could have no connection to reality. What I have tried to do is show the possible, underlying cause of a phenomenon, using the simplest, physical explanations. I believe there is an ultimate reality that should be sought.

SPACE-TIME

All that the cosmos consists of is matter, energy, space and time. I believe that the first three of these things are fundamental and that the last (time) is not.

All of the particles and bodies in space are matter. Energy when applied to matter causes it to accelerate and achieve an increased velocity. Space is the distances between bodies and can be assumed to extend beyond the most distant body.

Time is determined by making use of the above statements. Time is the duration of the motion that a body undergoes at a given velocity to travel a distance in space. It is given by the ratio of distance to velocity. Distance and velocity are observable, independent variables. Time is the dependent variable.

We can see two objects at a distance apart. We can move from one object to the other, slowly or quickly. We cannot see time. We can use the duration of the day to give us a time reference. We can walk, putting one foot against the other for 5280 steps and define this distance as a mile. We can walk slowly, back and forth along our mile, and count the number of times we do so, for a whole day. The number would be our velocity in miles per day. If we walked faster the number would be higher and we would calculate a higher velocity.

We can mechanically or electronically divided the day into smaller parts, hours, minutes, seconds, and microseconds. We can use these to express the ratio of distance to velocity, for very short distances or very high velocities. Sixty miles an hour is still the same velocity as a mile a minute.

This is a key concept when we consider the travel of light over a distance. The velocity of light has been defined by Einstein to be a universal constant, the maximum velocity physically achievable. The vast distances in space are described in units called "light years". (These years are Earth years, not Mars years).

The source of ones perception of time is the brain. Within the brain, the distance between two adjacent nerve endings and the velocity of the neural transmitter molecule that travel across this distance determine the quickness of ones perceptions and reactions. The basic clock rate of our thought process depends on distance and velocity.

Einstein developed his theories of relativity on the basis of velocity and time being independent variables, which made distance the dependant variable. His four dimensions are the three spatial dimensions (X, Y and Z) and a distance dimension determined by the product of velocity and time. This ultimately led to one of his conclusions that had time varying as a function of velocity relative to the speed of light, independent of distance. It also led to the concept that space itself could expand or contract. I believe that having time as an independent variable was one of the factors that led to these illogical results.

My belief is that the observed motion of light and matter can be explained by their interactions with gravitons. It is gravitons that produce the effects of gravity. There is no need to be bound by Einstein's relativity theories. This allows space to be considered as invariant, infinite in extent. Also, this allows time to be considered as merely the infinite duration of the total of all events.

STARLIGHT

The Sun and all the stars we see with our naked eyes are part of what is called the Milky Way Galaxy. Our galaxy is composed of billions of stars arranged in a flattened disc shape. Our Sun is located about half way between the center and the edge of the galaxy. When we on Earth look toward the center of the galaxy the stars appear to get closer and closer together, until they seem to become a solid band of white, the Milky Way. If we were to look between the stars through a large telescope into the far reaches of the universe we would see many other galaxies like the Milky Way. There may be billions of them.

The measurement of the wavelength of light is called spectroscopy. It has been very important in the understanding of stars, galaxies and the cosmos. Many wavelengths are mixed together in the light coming from the Sun. The early scientists that studied this sorted them out with some difficulty. Laboratory tests under controlled conditions helped. After establishing the nature of these spectra they were able to detect the shift that had occurred in the spectra from the light of distant stars and galaxies.

Measuring the spectra of the light emitted has led to the knowledge of star size, temperature, and direction and speed of motion. The hotter the star the higher the energy of the excited hydrogen in its atmosphere and therefore the bluer its radiated light due to the higher energy photons.

It should be noted that there are two kinds of spectra. One consists of the bright lines emitted by the excited atoms. The other spectrum consists of dark lines in the almost continuous bright light from the star. This spectrum is produced when cooler atoms, selectively absorb photons passing through the outer region of the star.

Color is an important characteristic of light. It is what we perceive in the visible portion of the spectrum of light. But we should consider it as the wavelength of light, whether it is infrared, at a longer wavelength than visible light, or ultra-violet light, at a shorter wavelength than visible light. As an electro-magnetic wave, a specific "color" of light oscillates at a specific frequency as it travels along at light speed. The distance that it travels in one cycle of oscillation is the wavelength.

As an example, the wavelength of red light is almost a micron, which is a millionth of a meter. Since the speed of light is about 300 million meters per second and the frequency is the speed divided by the wavelength, this gives a frequency for red light of about 300 trillion cycles per second. This is why light is usually characterized by its wavelength.

An important phenomenon to understand is what is called the "red shift". This is used to measure the velocity of a star or galaxy relative to the Earth. The greater the velocity of separation, the more the characteristic blue light appears to be shifted towards the red end of the spectrum.

To understand this, consider a star moving away from us at a tenth the speed of light. Think of the first cycle of the light at the beginning of one second, moving the known distance of 300 million meters during that time. The last cycle, emitted after the star had traveled one second, would be 330 million meters behind the first cycle. Dividing this slightly greater distance by the known frequency of the light results in a slightly longer wavelength. Therefore, when we detect this radiation, it appears shifted towards the red end of the spectrum.

Another aspect of this is that the same effect would occur if the distant object were stationary and we were moving away from it at some fraction of the speed of light. During a one second period, that we would measure the light, slightly less of the radiated wave would pass us. This would result in a lower number of cycles per second being detected. This results in a redder, longer wavelength light observed.

UNIVERSE

Observations have been made of the light from quasars, estimated as having been emitted about 12 billion years ago. These emissions are thought to have taken place only one billion years after the Big Bang. Repeated observations of these objects indicated that the rate of change of their red shift was increasing. It was concluded that the expansion of the universe was accelerating. A negative force of gravity was postulated to account for this occurrence.

However, studies of the distribution of galaxies, as a function of distance from us, indicate that space is essentially flat. I will take a Newtonian view of the universe and assume space is linear and not expanding. I believe there is a different way of explaining an increasing red shift in the above example.

Instead of the usual assumption that all galaxies are moving away from us with velocities that increase with distance, assume that astronomical bodies move away from where the Big Bang occurred with a wide range of radial velocities. Also, assume that the Milky Way galaxy has about an average radial velocity.

The quasars may have been moving very slowly, either toward us or away when they emitted their light. If they had a decelerating outward motion or an accelerating inward motion, the result would be an increasing rate of red shift change.

The amount of red shift of light we observe from distant objects depends on the rate of increase of the distance between them and our galaxy. There are several ways that we may observe the same degree of red shift.

The distant source of light could be moving in a direction opposite to the motion of our galaxy. The observed red shift would be a function of the sum of their velocities.

There could be an object moving in the same direction as our galaxy, but much slower. It would then be closer to the center of the universe than our galaxy. Another object could have started in the same direction as our galaxy but with a greater radial velocity. It would then be farther away from the center than our galaxy. In these cases the red shift would be directly proportional to the difference in velocities between them and us. The following figure illustrates these things.

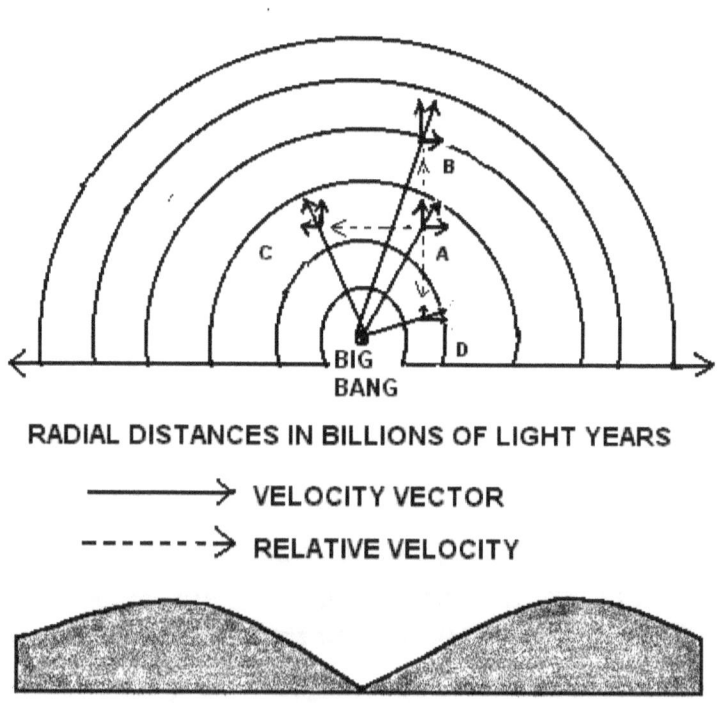

RADIAL DISTANCES IN BILLIONS OF LIGHT YEARS

$\longrightarrow$ VELOCITY VECTOR

$------\rightarrow$ RELATIVE VELOCITY

GALAXY POPULATION DENSITY (CROSS SECTION)

Figure 1

Assume that our galaxy is located at point A. Its radial velocity is shown resolved into vertical and horizontal components. Point A is shown at a distance from the center (The Big Bang) that is near the peak of the galaxy population density

A galaxy at point B had started with a larger radial velocity than our galaxy. The vertical velocity component of the B galaxy was greater than that of our galaxy at the time that B emitted the light that we now receive. The red shift that we observe is a result of the difference between the two vertical velocities.

The galaxy at point D started out with a smaller radial velocity that our galaxy. Its vertical velocity was less than that of our galaxy. We were moving away from D at the time D emitted the light we are just detecting, with the observed red shift as a result.

Another possibility is to consider a galaxy that started with about the same radial velocities as our galaxy but at different azimuth angle. This is shown in the above figure as the galaxy at point C. This galaxy and ours are both moving away from each other. The red shift results from the sum of the horizontal velocities. This red shift would be considerably greater than for the other two cases, even though all are at the same distance from us.

This could account for the almost two to one variations in the measurements of the Hubble Coefficient by different observers and the various estimates for the age of the universe of from 10 billion to 20 billion years.

It is reasonable to assume that other universes preceded ours. While the course of the expansions and contraction of the prior universes cannot be known precisely, we can assume that they ended with conditions that led to the initiation of the next universe. It is likely that there was sufficient matter that converged to a central concentration of mass that produced a "Big Bang".

If a model of an expanding and contracting universe were made it might show that by the time an appreciable amount of matter had accumulated at a central point, much of the remaining matter would be relatively close, consisting mostly of galaxy-weight black holes.

Assuming that the concentration of exotic matter in the central mass had reached a critical density or state, an impact from another massive black holes may have triggered the Big Bang. I believe that the Big Bang was a fact of nature that happened as a naturally occurring event. Also, if something happened once, then the conditions that led to it can happen again.

When the detonation occurred and the expanding plasma reached the point of atomic particle formation, they could have started responding to the mass of the surrounding black holes. Some of the high velocity plasma would have been trapped into orbits around the black holes, perhaps producing what we detect as quasars. The disruptions in the outward flow of the plasma would have resulted in variations of density and velocity. This could account for the eventual clusters of galaxies and the voids between galactic clusters.

The momentum of the expanding plasma would interact with the momentum of inward moving black holes. The results could be a slowing of the inward motion or a reversal of the motion of the black holes. The massive black holes would probably resume their inward motion with some exhibiting quasar activities.

A black hole with only the mass of a single star would be capable of attracting a concentration of expanding gasses but would be carried along with the gasses. This might explain the initiation of galaxies with black holes at their center.

MULTI-VERSE

The following is another possible explanation for the change in the rate of red shift for the most distant objects we observe. Massive objects lying beyond our universe perhaps could affect the outermost objects in our universe. Perhaps other universes surround our universe.

The prefix "uni" in universe comes from the Latin for "one" but in the word "universe" it is considered to mean "the one and only". I believe it is conceivable that our universe may be only one of many simultaneous universes. I would like to call this the "multi-verse" consisting of all the universes, since "multi" is the prefix for "many". This is my understanding of what the "cosmos" can be considered.

My ideas about simultaneous universes will need some explaining. I believe that space is endless, infinite in extent, and linearly three-dimensional. I also believe that time is also infinite in extent, without a beginning or end.

Assuming that the age of our universe is 20 billion years, at most, then the light emitted at the time of the Big Bang has now reached the surface of a sphere with a radius of 20 billion light years. The sphere containing any appreciable matter, such as high velocity gasses, probably has a slightly smaller radius.

We can now picture our expanding universe as an isolated sphere floating in infinite space. What is wrong with this picture?

Why should there be only one unique universe? An infinite space has room for an infinite number of universes.

I believe that the existence of any matter and energy at all, means that there can be an infinite amount of matter and energy in existence.

I have pondered on the source of all of this and have come to a conclusion. If one believes that everything was divinely created, then that is the answer. The only alternative that I can see is that all matter and energy has always existed. I often come back to thinking; "but where did it all come from originally?" I have to remind myself that it always existed.

If there are other universes surrounding ours, how could they be spaced around ours? With the simplifying assumption that they are all spherical and about the same size, the densest configuration would be for about twelve other universes to surround ours, essentially in contact. This would place ours in a somewhat favored position, which I don't think is warranted.

Or other universes could be located randomly around ours at varying distances and orientations. I think that over a large number of iterations, of interactions and big bangs, the random configuration would come to have a fairly regular arrangement.

A simple hypothetical assumption would be to place our universe in a cubic lattice. This configuration could be repeated indefinitely in all directions with none of the universes in a favored position.

We could picture our universe at the central vertex of the eight surrounding cubic volumes. This closest cubic array has a total of twenty-six other universes. This is shown in the following figure.

Let us consider the dimension of the side of the cube to be some unit to be determined.

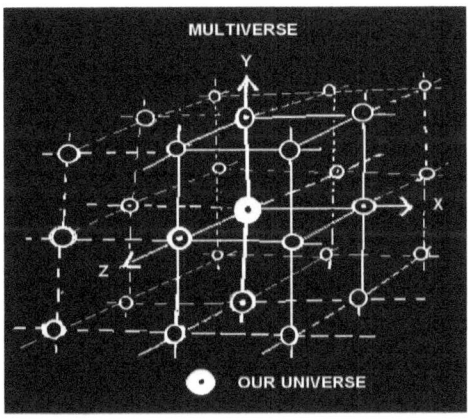

Figure 2

There would be six universes at this closest distance away from ours. These six universes are those along the X, Y and Z-axes. There would be twelve more universes centered at a slightly larger distance away, the square root of two times the unit distance. Finally, there are eight more centered at the square root of three times the unit away.

The gravitational effect of these surrounding universes should be calculable. At the center of our universe they will produce equal and opposing fields from each opposing pair of universes. This effect could represent what has been proposed as the Higgs field.

As an outermost galaxy in our expanding universe gets closer to one of the other universes and farther away from the opposite universe, it may be effected by a difference in the gravitational effects, and undergo acceleration toward the closer, adjacent universe.

From the acceleration that has been observed, an estimate of that unit of distance between universes might be attained. (The above assumed that gravity is an attracting force, which I no longer believe.)

Although the surrounding universes may be at different stages of expansion they eventually could reach a point were two or more of them start to overlap. This might produce a situation where in that region the concentration of mass would reach a larger density and start to compress toward a new "Big Bang". Obviously this event would occur in the regions between groups of universes. In this way, universes that are expanding towards oblivion can give birth to new universes by interacting with one another.

That all the universes would begin at about the same time is based on the following assumption. Any expanding universe would spread an eighth of its total mass into each of the adjacent lattice cube volumes. Each of the universes at the eight corners of each lattice cube would do the same. The largest concentrations of mass would eventually occur in the center of each cube. This is where the total mass would again reach the critical value. The following figure shows eight universes expanding toward each other.

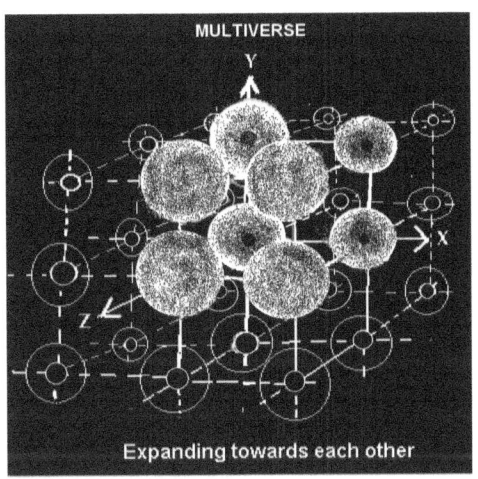

Figure 3

Groups of two or three of these universes can start accumulating large combined black holes where they overlap. Eventually all of these can collide in the center and trigger the Big Bang.

GRAVITY

Newton assumed that gravity is a force of attraction acting between material bodies. His approach to gravity is appealing for its simplicity but does not address what causes gravity. It only shows how the effects of gravity can be calculated.

Einstein, in his development of the theory of general relativity, also assumed that material bodies exerted a force of attraction that affected the space-time continuum. Neither man explained why this attractive force occurred. Einstein's mathematical approach leads to the concept of curved space-time causing the effects of gravity. By manipulating equations involving multiple variables and coefficients he was able to predict known experimental results. His explanation for gravity was that objects followed paths of least resistance in curved space-time. As a result of his equations, some explain the expanding universe merely as space expanding.

I have been struck by the need to have gravity act over great distances essentially instantaneously. I think that the gravity of the total mass of our universe may be felt far beyond the present radius of the Big Bang radiation.

What if at the time of the Big Bang, when the initial state consisted only of gravitational force, a gravity wave was radiated at a velocity far in excess of the present speed of light? What if gravity is always transmitted faster than light? Would it be an attracting force or a repelling force?

In my concept of multiple universes, I assumed that all the universes surrounding ours began at about the same time. All of these universes would have about the same total mass as each other. This is based on my concept of an accumulation of giant black holes reaching a critical mass and creating a Big Bang.

The gravity waves from all of these universes would be permeating our universe from all directions. The effect of this would tend to balance out on the large scale within our universe. However, if these gravity waves were repulsive, a large body could tend to shield a smaller nearby object from the waves coming from beyond the larger body. This would make the smaller body move towards the larger.

One theory of gravity assumes that gravity is transmitted by an intermediary particle called a graviton. It is said to be a mass-less particle traveling at the speed of light and having momentum. If gravity is transmitted by a mass-less particle, then there is no reason that at would be limited at the speed of light.

Now I am proposing that gravity is caused by gravitons transferring momentum to the atoms with which they collide. The greater the number of protons, neutrons and electrons in a body per unit volume, then the more the gravitons will interact with the body. I am assuming that these will be elastic collisions and not reduce the energy of the gravitons significantly. Many gravitons will pass through most material objects without any collisions. Those involved in collisions will be scattered over a range of angles. Any isolated material body in space would be subjected to equal amounts of momentum from gravitons coming from all directions of space.

Therefore, there would be no net momentum added in any one direction. However, if a larger body were in the vicinity of a smaller body, it would shield the smaller body from receiving some of the gravitons coming from the direction beyond the larger body. This is shown in the figure below.

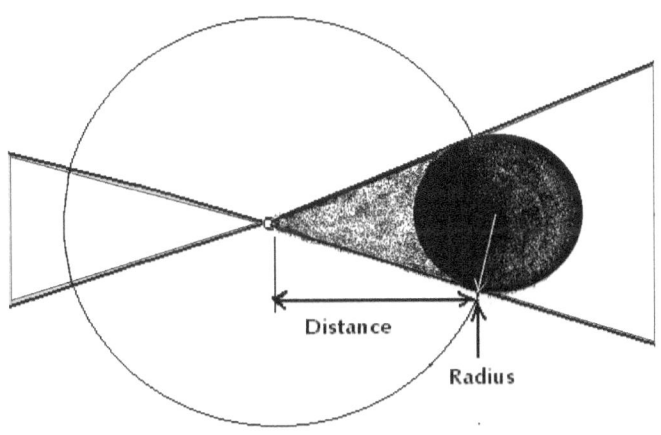

Figure 4

The amount of shielding would depend on the presented area of the larger body and its distance from the smaller one. It also depends on the density of the large body and how many of the gravitons it intercepts and how many it transmits.

Through the same solid angle that the large body presents, but in the opposite direction from the small body, the incoming gravitons are unimpeded. This results in a net difference in momentum applied to the small body. This will cause it to move towards the larger body with a specific acceleration.

The effect of the momentum will depend on the mass of the small body. For a given volume of matter, the greater the density, the greater the mass, and the greater the force applied to the body by the gravitons. Since force is equal to mass times acceleration, if the ratio of force to mass is constant, then the acceleration is constant. This is consistent with the gravitational coefficient of Newton.

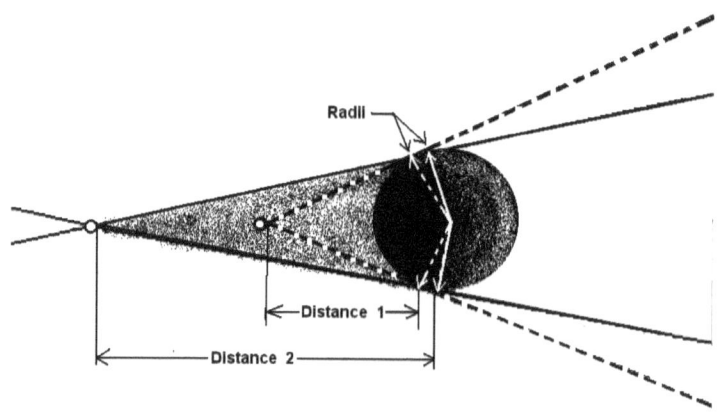

Figure 5

The solid angle presented by the larger body depends on the distance between the bodies. As shown in Figure 5, the solid angle will vary as the inverse square of the distance.

The inverse square relationship can be explained as follows: The presented area of the larger body is approximately pi times its radius squared. This will be a small fraction of the total area of a sphere centered on the smaller body, calculated for the specific distance to the large body. The area of this sphere equals four pi times the distance squared.

The fraction of the total radiation intercepted by the larger body is given by the ratio of these expressions. This is the radius of the larger body squared, divided by four times the distance squared. This simplifies to a constant divided by the distance squared.

This applies actually when the presented angle is small. As the smaller body approaches the larger body, this angle will become large and the relationship with distance will deviate from the inverse square.

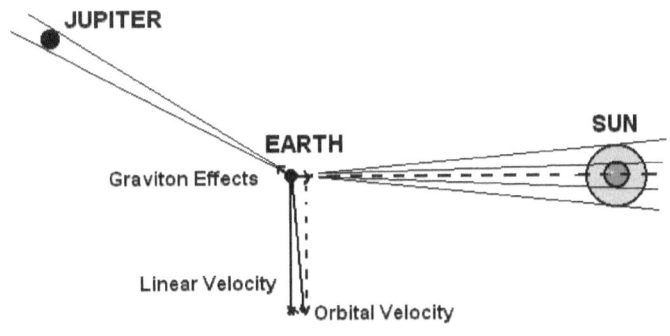

Figure 6

In the figure above an approach to calculating the effect of the Sun on the orbit of the Earth may be considered. The momentum applied to the Earth to change its linear velocity direction into its orbital velocity direction may be calculated. If the scattering coefficient of gravitons, for the various regions of the Sun and the Earth, can be determined, the net reduction of graviton momentum applied to the Earth could be calculated by integrating over their full volumes.

Similarly, the effect of Jupiter, at any particular distance and relative orientation might be calculated. The known deviation in the Earth orbit caused by Jupiter can be accounted for by the decrease in graviton flux caused by Jupiter.

Knowing the total mass of the Earth and the force necessary to accelerate it into a circular orbit should allow a calculation of the effect of the graviton flux coming from the opposite directions of either the Sun or Jupiter.

WEIGHT

Consider a small object that was caused to approach a large sphere until it stopped in contact with the sphere. This is shown in the following figure. The momentum applied to the small object will now cause it to press against the large object with a force that results in what we measure as its weight.

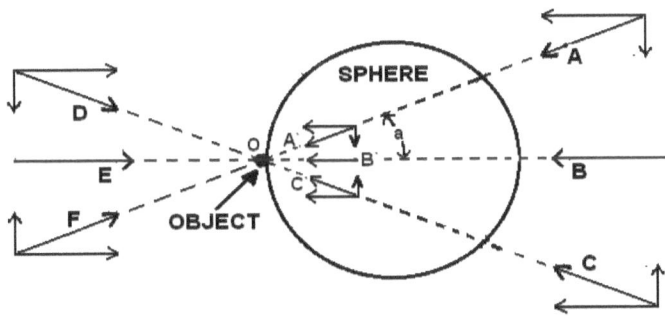

Figure 7

Assume that the sphere is the Earth. Any graviton passing through the small object from the right will have gone through the Earth for various distances. The horizontal path labeled B is the longest distance, through the diameter of the sphere. This path will cause the most reduction in the graviton flux. However, all of the exiting flux, labeled B`, will be along the horizontal direction and affect the measured weight.

For every path such as A there is an equal length path, such as C. Their path lengths will be shorter than the path B and will not reduce the graviton flux as much.

Their horizontal components will be reduced by a factor equal to the cosine of angle a. These horizontal components will add. The vertical components of A and C will be equal but oppositely directed and will cancel.

The horizontal components of the un-attenuated graviton fluxes, such as D, E and F, will add in the object O, and will be partially cancelled by the sum of the horizontal components of all the attenuated flux entering the object from the right.

The fundamental factors that will have to be determined for calculating weight are the following:

1. What is the un-attenuated, graviton flux density, coming from all direction of space?
2. What are the attenuation factors for various large bodies such as the Earth and the Sun? (i.e. Scattering coefficients, absorption coefficients, etc.)
3. How much force is transferred to the small body by the differential graviton flux?

In Figure 8 below, an approach to begin the calculation is shown. Simple trigonometry and calculus can be used to solve the flux components as a function of the angle "a". For a flux path through point B and going through point O, the distance "x", is given by the cosine of "a" times the diameter "d". This is because all triangles subscribed in a semicircle are right triangles.

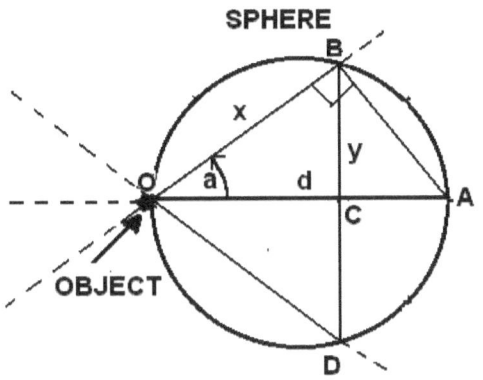

Figure 8

The distance "y" going from point B to point C, is given by "x" times the sine of "a".
Therefore: y = d (sin a)(cos a).

All the paths that have the same value of angle a, enter the sphere through a circle whose circumference equals pi times 2y.

Each path through the sphere will have an attenuation factor that will vary as a function of the length of x. This function remains to be determined.

Put all of these together and add another factor of cos a, to convert to the horizontal components of the flux. Integrating the resulting expression from a = 0 to 90 degrees should give the total, attenuated horizontal component of flux entering the small object from the right.

Similar equations can be used to arrive at the total horizontal flux entering the small object from the left. Obviously, the attenuation factor can be omitted in this case.

An experimental start to determining some of the unknown parameters might be to measure the weight of a mono-crystalline object, with a simple geometry, such that its mass and total number of atoms can be calculated. Depending on the structure of the crystal, there may be differing probabilities of graviton interactions with its atoms, under various orientations.

From the probability of a graviton encountering an atom over the range of possible angles of incident, the average force that is transmitted to an atom might be obtained. Of course the ratio of the flux coming from the sky relative to that coming through the Earth must be determined.

On further thought, for a very small mass, the path length through it would be very short. Since this mass exhibits weight, the probability that gravitons interact with it is 100%. This means that the probability of a graviton passing through the Earth and not interacting with an atom is close to zero. This suggests that this attenuated graviton flux passing through the Earth may be neglected in the calculation.

There may be a minimum time duration required for sufficient gravitons to interact with a small body and arrive at a stable weight reading.

LIGHT

Trying to decide the nature of light has been a quandary for many centuries. In some ways photons of light behave like particles and in other ways they act like waves. For a hundred years it was assumed that light, as a wave, needed a medium in which to travel through space. This medium was called the "luminous ether."

A problem arose in trying to measure the velocity of light. It had been assumed that the velocity of earth traveling through the ether would produce a change in the speed of the light moving in the direction of the motion of the earth, as compared to the speed measured at right angles to this direction. Many attempts were made to detect this change in velocity, using instruments with ever increasing accuracy. No change in the velocity could be measured.

In a last effort to retain the concept of ether as the medium for electro-magnetic radiation, a theory was proposed that the units of length and time, used to measure the velocity of light, must change as a function of the velocity of light. The equations that described this effect are called Lorentz Transformations.

Einstein used these equations in his "Special Theory of Relativity". He applied them to all matter, except for light. He defined the velocity of light as a constant that is the maximum velocity that can occur. He assumed that this velocity was independent of the velocity of the source of the light. This was his attempt to reconcile theory with physical measurements.

Einstein expanded his use of the Lorenz Transformations. He stated that an accelerating object would have its mass approach infinity, as its velocity approaches that of light. Therefore, by this definition, a photon of light could have no mass.

This is in conflict with the known fact that light is able to impart momentum. Einstein knew this. His first paper, on the photoelectric effect, described how photons of light liberate electrons from the surface of a metal.

I assume that photons of light have mass. I also assume that the velocity of light is not the basic limit of the velocity of matter. Something else must be limiting everything, including light, to their observed velocities and masses. This led me to a new theory of gravity that I describe in this book. My concept of gravity is that it is caused by particles I call gravitons, traveling at "light speed." This has a direct bearing on my concept of photons of light being particles, having a finite mass, traveling at slightly less than "light speed".

One can visualize a photon traveling and being retarded by gravitons coming from the direction ahead. The photon would also be propelled by gravitons coming from its rear. If the photon reaches the velocity of the gravitons following it, the photon will no longer be propelled. Then the gravitons coming from the front will retard the photon until a balance is restored. This should occur at a velocity just below the velocity of the gravitons.

This process could produce an oscillation of the photon, back and forth along its direction of travel. Oblique interactions with gravitons could cause side-to-side motions of the photon.

The energy of a photon and the equivalent mass of that energy can be calculated. To derive the actual parameters of a photon, start with the equation for a quantum of light. The energy of the photon, E is given by the following equation: $E = hn$ (joules)

h is Planck's constant.

$h = 6.626 \times 10^{-34}$ (joule-seconds)

n is the frequency of the radiation in cycles per second, given by the equation: $n = c/L$

c is the speed of light ($c = 3 \times 10^8$ meters per second)

L is the wavelength of the light in meters.

Therefore: (Equation 1) $E = hc / L$ (joules)

Assume that the mass of the photon can be determined by using Einstein's equation:

(Equation 2) $E = mc^2$. (m is mass in kg)

Combining this with Equation 1 gives the following:

$mc^2 = hc/L$

Therefore: (Equation 3) $m = h/cL = (2.2 \times 10^{-42})/L$ (kg)

For green light, $L = 5 \times 10^{-7}$ (meters). Therefore, the mass of a green photon is about 4×10^{-36} (kg).

By a different calculation the mass of an electron can be shown to be about 10^{-30} (kg). Therefore, the mass of a green photon is about four millionth of the mass of an electron.

It seems reasonable that an excited atom could throw off such a small portion of an electron mass and charge. Traveling at the characteristic speed of light, it will provide the observed momentum of the photon.

For a possible explanation of the electro-magnetic characteristic of the photon, assume the following: In addition to its forward motion, the tiny charged particle vibrates in a plane transverse to the direction of propagation

The moving charge will induce a magnetic field that will tend to stop its transverse motion at the end of its excursion. The magnetic field will then start to collapse and cause the charged particle to reverse its sideways motion.

As the magnetic field reaches zero, the particle will have reached its maximum transverse velocity and start increasing the magnetic field. This will be of the opposite polarity, which should cause the particle to slow down and stop at the other end of its excursion. Thus, the cyclic transfer of energy between the charge and the magnetic field will continue. This is shown in Figure 9.

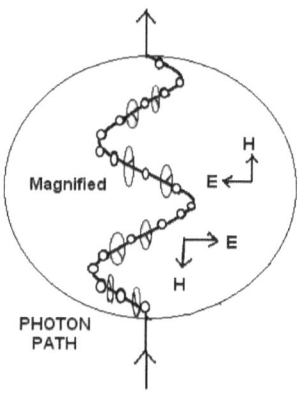

Figure 9

It is known that a photon with a higher frequency of oscillation has greater energy content. This suggests that a higher energy photon, with a larger mass and charge, will vibrate faster, with a smaller excursion.

It seems possible that the direction of the vibration could rotate around the direction of travel. This would account for circularly polarized light.

A characteristic of electrons that should be considered is that of electron spin. The spin of an electron around an axis through its center produces a circulating charge that causes a magnetic field that emerges at one end of the axis and returns at the other end.

If the small fraction of an electron that is expelled as a photon retains a portion of the spin, this could account for the magnetic portion of the electro-magnetic field of a photon. Its electric field should be oriented transverse to the magnetic field. This is shown in Figure 10.

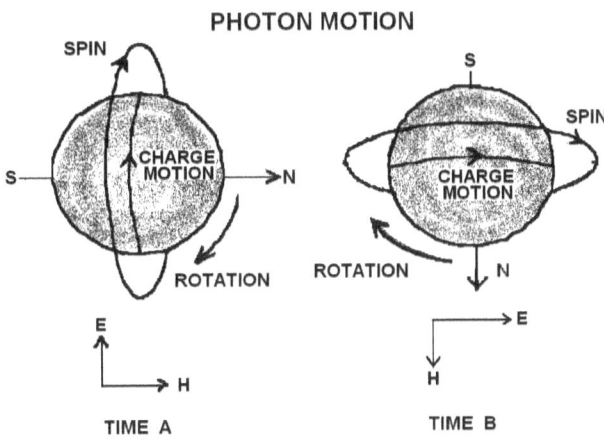

Figure 10

Figure 9 is a representation of the zero spin possibility. Figure 10 shows how the spinning charge might produce the electro- magnetic fields.

The direction of the charged photon path, in Figure 10, can be either across the page or into the page. A quarter turn of rotation is shown.

The fact that the polarity of the electro-magnetic wave reverses every half wave- length that the radiation travels could be accounted for by the particle axis rotating in a plane transverse to the direction of motion.

Another possibility is that the axis of spin could be oriented along the direction of motion. If the axis of spin were to rotate in the plane through the direction of motion, the wave front of the magnetic field would change polarity every half turn and go through a null at the time of reversal.

If the electric field were at a maximum at that instant, when the axis of spin is pointing along the direction of motion, the required alternating maximums of the electro-magnetic field might result.

I have described several possible modes of the photon particle motion to result in an electromagnetic field. I have learned that the value of spin assigned to an electron is ½. The photon has been assigned spins of 0 and ½. This might mean that both the modes described, that assumed spin and no spin photons, could apply.

RADIO

It has been generally accepted that all electromagnetic waves are the same. They are suppose to be described by the Maxwell equations equally well for radio waves and light rays. How can my concept about photons be compatible with radio transmission and reception?

When an alternating voltage signal is applied to a dipole antenna, shown in Figure 11, the following events are said to occur. I will assume a typical broadcast frequency of one hundred megahertz. A half wave dipole would be about one and a half meters long.

A sine wave carrier is applied to the middle of the antenna as shown. At the maximum of the applied power the electron flow from one side of the dipole to the other will reach its peak value. A maximum magnetic field will be developed circling the current flow. This field will move away from the dipole at the speed of light.

As the applied power decreases and goes through zero, the electron current flow stops and no magnetic field is formed. At that instant the maximum number of electrons have accumulated at one end of the dipole and the corresponding number of positive charges have appeared at the other end. That is where the electrons have been depleted by the current flow.

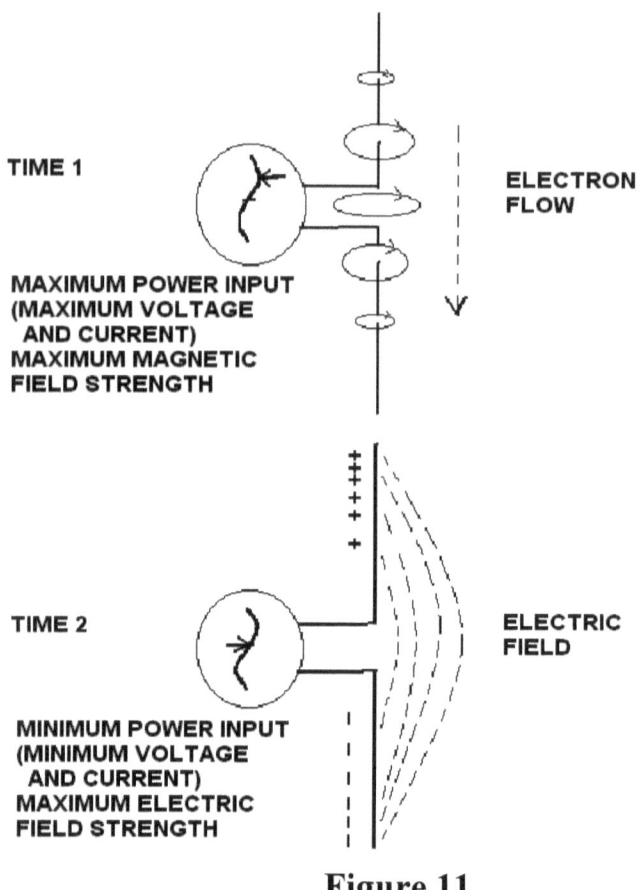

TIME 1

ELECTRON
FLOW

MAXIMUM POWER INPUT
(MAXIMUM VOLTAGE
  AND CURRENT)
MAXIMUM MAGNETIC
FIELD STRENGTH

TIME 2

ELECTRIC
FIELD

MINIMUM POWER INPUT
(MINIMUM VOLTAGE
  AND CURRENT)
MAXIMUM ELECTRIC
FIELD STRENGTH

Figure 11

As shown above, a shell of electric field lines suggests a connection from one side of the dipole to the other. This electric field also moves away from the antenna at the speed of light. It is obvious that in the next quarter of the cycle of voltage the electron flow will reach a maximum value going in the opposite direction. This will develop a magnet field circling in the opposite direction. This field will move off following the electric field, etcetera.

Electric power is put into the antenna and power is transmitted. The antenna is made of a good electrical conductor so that not much power is wasted in the electron flow, back and forth. It has been assumed that the excess input power is carried off in the electromagnetic fields. But what if there is the equivalent of photons that carry the power? What if the electric field lines consist of a line of tiny bits of electric charge emitted as the electrons move along the antenna? The bits would have a small motion along the field line. But their primary motion would be away from the antenna "at the speed of light."

There would be an elemental magnetic field ahead of them and another magnetic field of the opposite polarity developing behind them. After that magnetic field would come more tiny bits of charge with a slight motion in the opposite direction from that of the first particles.

In Figure 12 there is shown several of the cells, of the many, that form along the length of the antenna. These would interact in a complex interlocking circular motion while traveling outward.

The ovals represent the circulating magnetic flux. The electric charges move only slightly sideways, as they are moving outward at the speed of light. The pairs of charges are linked by their magnetic flux, as they tend to move tangentially past each other. The fluxes add between pairs and produce the alternating magnetic fields of the radiation (indicated by H).

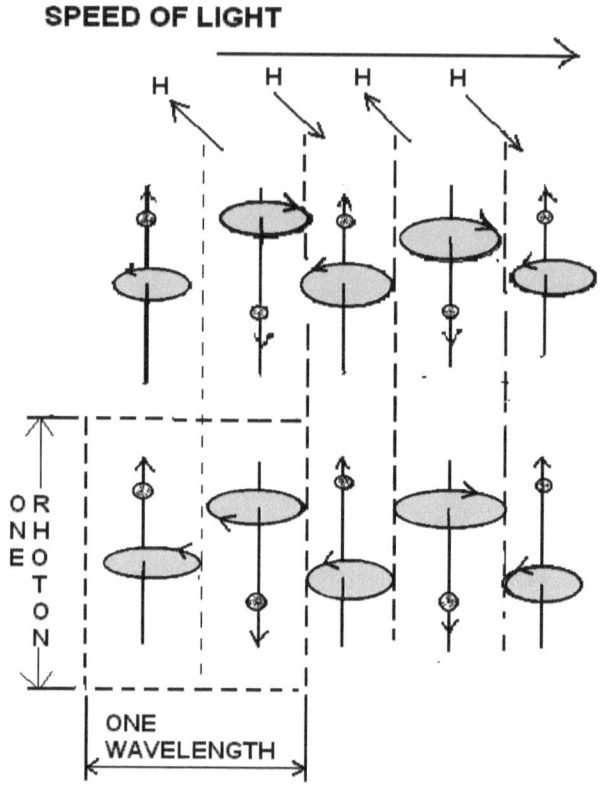

Figure 12

Here I go inventing a new term. If there are radio photons, let us call them "rhotons." They consist of two charge bits per wavelength.

The power from this antenna is radiated in a pattern shown in Figure 13. This is a vertical cross section of the field strength of the indicated dipole. This pattern is toroidal but has also been called doughnut shaped. Looking down at it would show a circular pattern.

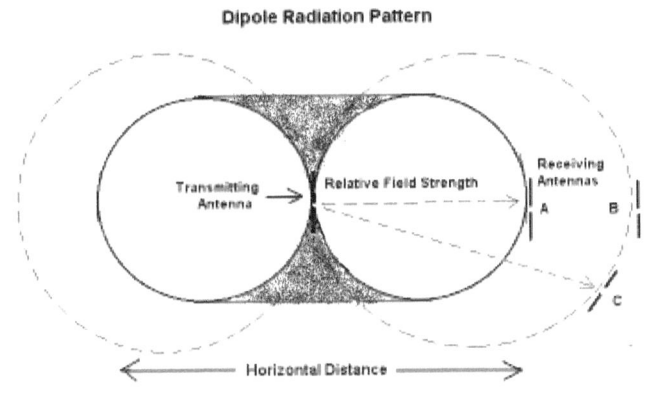

Figure 13

Three different receiving dipoles are shown at different distances and orientation relative to the transmitting dipole. Receivers at A and B are along the line of maximum field strength. B, being farther away than A, is at a point of lower strength. The receiver at C is off the maximum direction and has the same field strength as at point B, but at a closer distance to the transmitting antenna.

It is known that the field strength falls off with distance. With our assumption of elemental cells of oscillating charge and magnetic flux, how can the reduction in strength be explained? With the increasing area that the power is spread over with increased distance, the number of these cells in a unit area will decrease.

If the transmitting antenna is emitting radio type photons, what can be the mechanism? It is known that the electron flow on a transmitting antenna is mostly along the surface. This is called the skin effect.

Electrons moving in a metallic conductor, under the influence of an applied voltage potential, migrate by freeing valence electrons and by filling vacant valence sites. Perhaps this low energy transfer process would result in low energy quanta emitted from the surface atoms.

The common metals used in making antennas can be those such as steel or copper. They could be coated with silver. The atomic structure of these metals have from three to four completed electron shells surrounding the nuclei, with only one or two valence electrons in the last incomplete shell. The positive electric field of a nucleus is largely shielded from an incoming electron. So an electron interacts with the atom with a small velocity and a little energy exchange.

If the rhotons of the radio wave are comparable to photons, their characteristics should be calculable. Assume that Planck's equation,
$E = hn$, can be used to determine their energy. The 100 MH signal ($n=10^8$), would have an energy per rhoton of 6.6×10^{-26} joule.

Assume the transmitted signal power is 100 watts. That is 100 joules per second. One cycle of the broadcast signal takes 10^{-8} seconds. During one cycle of transmission, the amount of energy transmitted is 10^{-6} joules. The number of rhotons created in this period of time is 10^{-6} divided by 6.6×10^{-26}. This comes out to be 1.5×10^{19} rhotons.

To see how this output number relates to the power input to the antenna, consider that the radiation resistance of a dipole is about 70 ohms. The power input equals $I^2 \times R$, where I is the input current in amperes.

Solving for this current: gives the current as I=1.2 amperes. (An ampere is a coulomb per second.) A coulomb is a quantity of electric charge equivalent to 6.28×10^{18} electrons. Therefore, the input to the antenna is about 7×10^{18} electrons per second. During one cycle of input this is 7×10^{10} electrons. Dividing the above number of rhotons per cycle by this number of input electrons per cycle comes out to 2×10^8 rhotons per electron. This seemed like a lot to ask of each electron.

We will use the equation derived for the mass of a photon ($m=h/cL$) and apply it to a rhoton. In this case L=3 meters. Therefore, the mass of the rhoton would be 7×10^{-43} kg. Comparing this to the mass of an electron (10^{-30}kg), we get the ratio of 10^{12} rhotons per electron. The number of rhotons taken per electron is shown above as 2×10^8. The ratio of these two numbers is 2×10^{-4}. So the mass of the rhotons taken from an electron is only two ten-thousandth of the electron mass.

To get an idea of the density of rhotons in the radiated wave front, I made some simplifying assumptions. Referring to the Figure 13 above, showing a dipole radiation pattern, assume that point A is at a distance of one wavelength, 3 meters. The circumference of this vertical circle is about 10 meters. Most of the radiation will be passing out through the semi-circle, 5 meters around.

The diameter of the horizontal circle will average between 3 and 6 meters. I will assume 4.5 meters. Therefore, the circumference of this circle is about 15 meters.

The total area through which the radiation will pass is the product of these two lengths, 75 square meters. By inspecting the derivation of the this area it is obvious that the area will increase by the square of the distance. Therefore the rhoton density will fall off as the inverse square of the distance.

The rhoton density, at any specific distance, will be the number of rhotons transmitted per cycle, divided by the area they pass through, at that distance. As an example; the density at the distance of 3 meters would be 2×10^{17} rhotons per sq. meter. A receiving antenna, placed at that point, would intercept the number of rhotons that would fall on the presented area of the antenna.

If the antenna were 1.5 meters long by .01 meter wide, its area, of .015 square meters, would intercept 3×10^{15} rhotons per cycle. Assume that each of the rhotons, that impinged on the antenna, were to free a valence electron. The total number of electrons that would flow, under the influence of the accompanying magnetic fields, would be less than a thousandth of a coulomb. Hence, less than a milli-ampere of current would be detected. Since, after each half cycle of a rhoton, the direction of its magnetic field reverses, the above current would reverse its direction of flow each half cycle.

HYDROGEN

Almost everything we see in the sky is radiation from hot hydrogen atoms in the outer regions of stars. A very small amount of the light comes from the other trace elements. To understand how this happens we must consider the hydrogen atom.

The electron spins around the nucleus (a proton) at a relatively large distance, compared to the size of the nucleus. The atoms in the gas interact with each other more frequently with increased temperature. This raises the energy of the electrons in some of the atoms. The electrons return to their normal energy levels by emitting photons of light. It was found that the photons emitted from hydrogen atoms had a specific pattern of discrete wavelengths, as measured by a spectrometer. The radiation from other elements had their own distinctive patterns.

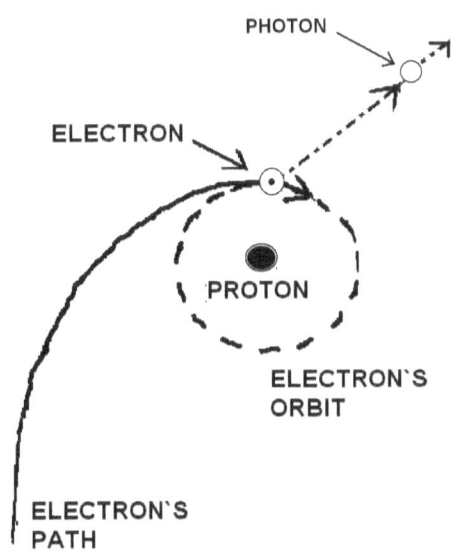

Figure 14

In Figure 14, the path of a free electron is shown as it approaches a positively charged hydrogen ion. As it nears the proton, the electron emits a photon of light and goes into orbit around the nucleus.

If this orbit happened to be the closest one to the proton that can occur, the photon emitted would have the largest momentum of any of the photons in the hydrogen spectrum. It would have a frequency that placed it in the ultra-violet region.

A theory about the hydrogen atom has been named The Bohr Model after the scientist who developed it. A representation of this is shown in Figure 15. This model has a number of nested orbits with systematically increasing radii. His theory states that electrons could only enter orbits with specific energy levels. The photons emitted would have momentum determined by the difference in energy levels between the starting and ending orbits.

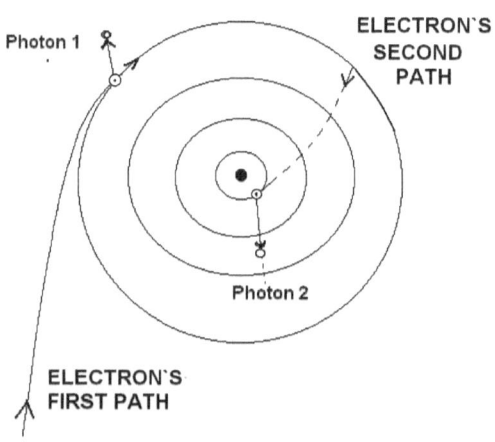

Figure 15

In Figure 15 the free electron first entered one of the outer orbits and the energy difference was small. Therefore, the first photon was of low momentum and frequency. The second transition shown was into the innermost orbit and had a larger energy change.

The second photon had a larger momentum and higher frequency than the first. Interestingly, the sum of these two photon energies equaled the energy of the photon shown in Figure 14.

Some electron transitions from various orbits going directly to the innermost orbit produce a spectrum with numerically related frequencies. This spectrum is called the Lyman spectrum after the scientist who discovered it.

Similarly, some transitions that end at the next closest orbit to the nucleus produce a spectrum named after its discoverer, Balmer. The highest frequency photons in this series are at the blue end of the visible spectrum.

Why does an electron go into orbit around a proton instead of continuing on and colliding with it? When I learned of the quark theory of the make up of the proton, I began to see why the electron acted as it did. Figure 16, shows an electron going into orbit around a proton. Obviously, the spacing between the electron and the proton is not to scale. I show the proton according to my concept of how it may be configured.

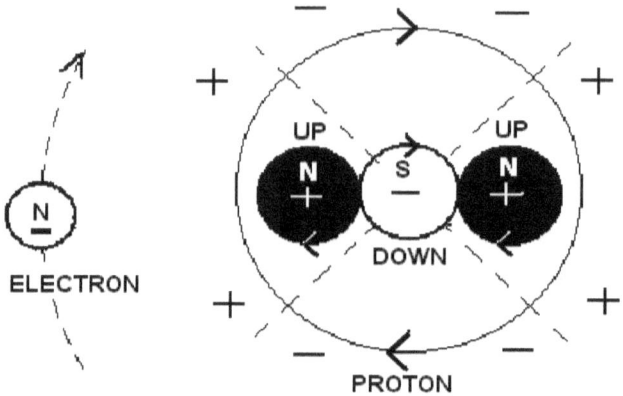

Figure 16

The proton has two Up quarks, each with a +2/3 charge. They would tend to repel each other. It has been stated that something called gluons keep them from flying apart. Instead of gluons I assume that the negatively charged Down quark between the Up quarks can tend to hold them together.

The Down quark, with its –1/3 charge, will attract the two Up quarks, somewhat. The total charge of the proton is the sum of all the quark charges, a positive one.

One trivial point bothered me. Why the –1/3 and 2/3 charges of the quarks? Then I realized that the definition of a unit of charge is completely arbitrary. We could define the Down quark as –1 and the Up quark as +2. Then the proton would be +3 and the electron would be –3.

94

In addition to their charges, the quarks have spin. This produces a magnetic field around each quark. Assuming that each quark spins in the same direction, the Up quarks will have their positive (north) magnetic poles in the same orientation and the Down quark will have its positive pole in the opposite orientation. This will also cause the magnetism of the Down quark to attract both Up quarks and hold them close.

The entire proton also spins. Its electric and magnetic fields will interact with the approaching electron as they traverse it. The electron spins and has its own electric and magnetic fields. It will be attracted to the Up quarks positive electric fields but be repelled by the same orientation of its magnetic field with that of the Up quarks.

As the Down quark interacts with the electron, the like electric fields will repel while their oppositely oriented magnetic fields will attract.

At some orbital distance these forces should produce a critical interaction. I believe that it is during this interaction that the photon is produced.

The nature of an electron has not been given much consideration. I believe an electron may have a granular or fluidic composition. Under what could be considered a tidal interaction with a nucleus, an electron may throw off small bits of itself (photons), to achieve an orbit around the nucleus.

Consider the various factors that are involved in the electron and proton interaction:

If the velocity of the electron were small compared to the rate of rotation of the proton, successive periods of electric attraction and repulsion would impinge on the electron at the rate of the proton spin.

If the velocity of the electron were a significant factor, the lengths of the alternating periods would depend on whether the electron moved in the direction of the rotation of the proton, or in the opposite direction.

To simplify the analysis, assume that the electron does not travel far during a short time interval of interest. The four different electro-magnetic zones of the proton will traverse the electron each revolution of the proton, for a number of revolutions during this time interval.

The electron is also rotating. At some combination of all the factors of the interaction, the electron could be swept by two successive fields of the proton, for every rotation of the electron. The effect of this is shown in Figure 17.

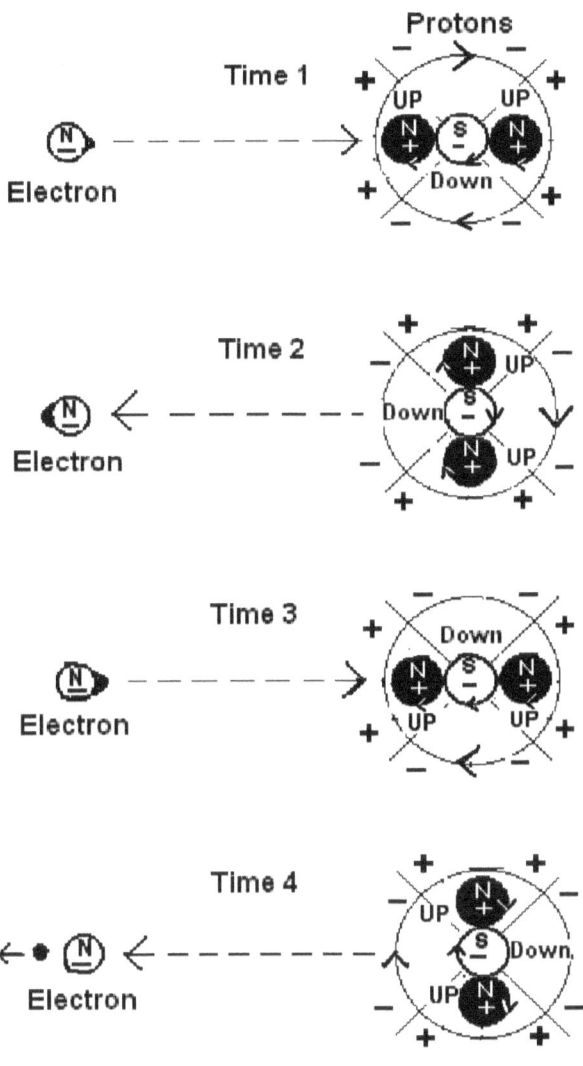

Figure 17

When the Up quark is facing the electron, its positive charge attracts the closest surface of the electron, raising a bulge. By the time that the Down quark is facing the electron, the bulge has been rotated to the opposite side of the electron.

The repulsion of the negative charge of the Down quark, further increases the size of the bulge. This cycle will be repeated until the bulge becomes large enough to separate from the electron and fly off as a photon.

The interaction of the magnetic fields, of the quarks and the electron, could produce motions of the entire electron that are synchronous with the above interaction. This additional motion could enhance the building of the bulge and the separation of the photon.

Assume the photon is launched from the far side of the electron, while the Down quark is repelling the charge, and as the magnetic field of the quark is attracting the electron. The contributors to the photon velocity should be the speed and spin of the electron, and the repelling, negative electric field. I am assuming that the interaction with gravitons will cause the photon to achieve the speed of light.

The questions remain, how does this interaction cause the electron to go into orbit around the proton? Does more than one of these bits of charge get thrown off until a stable interaction is achieved? Does a photon consist of a series of charges, perhaps with magnetic fields that alternate? If this were the case, the spacing between the bits could determine the wavelength of the photon.

The proton could rotate around an axis that was at right angles to the one in Figure 17. This is shown in Figure 18 below.

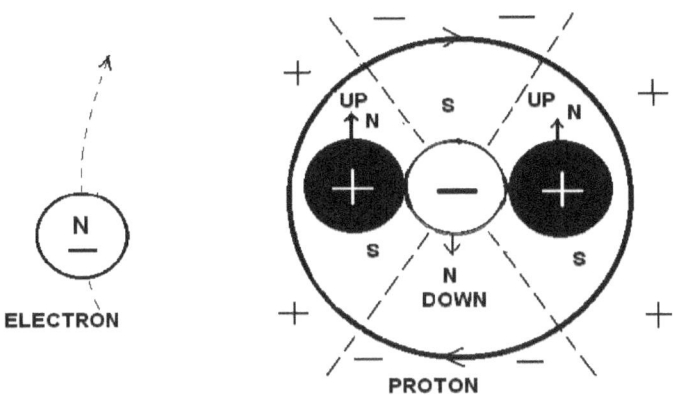

Figure 18

In this case, the magnetic fields of the quarks would not interact with the magnetic field of the electron, as shown. Only the electric fields of the quarks would alternately attract and repel the electron. This case might also tend to produce the bulge, and hence the photon. I think the contribution of the magnetic fields is needed.

I will assume that the spin rates of the electron and the proton do not change during this encounter. The main variables to consider are the velocity of the electron and the distance from the proton to the electron. I will also assume that the direction of the electron motion is the same as the direction of the sweep of the fields coming from the proton.

It seems that a stable orbit could come about when the electron is moving at the same velocity as the nucleus fields. The fields would no longer alternate across the electron and would not cause photons.

This would mean that the velocity of an electron in one of the larger, outer orbits would be greater than the velocity of one in a smaller, inner orbit. This is because each electron would have to make a single orbit in the time the proton rotates once. At any other electron velocity or distance, the effect of the alternating fields would depend on their rate of alternation, relative to the spin rate of the electron.

It can be seen in Figure 17, that two rotations of the electron are shown for every single rotation of the proton. Other odd multiples of this electron spin rate could also produce similar alternating interactions. Perhaps photons would occur with six rotations of the electron for each proton rotation, as well as for ten. Such a progression might occur as synchronous rotation is approached at different electron-proton spacing.

As a photon is emitted, the momentum imparted to the photon is subtracted from the momentum of the electron. This will tend to slow the electron. The electron will go to an orbit where it can revolve around the proton, in synchronism with the motion of the electric fields of the proton.

There are several conditions that could prevent an electron from going into orbit around a nucleus. The electron could approach too fast to be slowed sufficiently. It could be coming on the wrong side of the nucleus, and would have to circle against the sweep of the nuclear fields. The electron could come on the correct side but with the wrong direction of the electron spin polarity.

If an electron approached the nucleus from a direction along the nuclear spin axis, the nuclear fields would not affect it. It would respond to the net electric fields of the nucleus and the surrounding electrons. Such an electron would either be deflected or would continue on to impact the nucleus.

STARS

How does hydrogen, the smallest element, create stars? An electron spins around a single proton, the nucleus of the hydrogen atom. As previously shown, the proton is made up of even smaller components called quarks. It contains two "Up" quarks, each with a charge of +2/3. It also contains one "Down" quark with a charge of –1/3.

Figure 19 below, shows how a very high-energy electron can interact with a proton and produce a neutron. The electron can be thought of as having combined with one of the Up quarks, changing it to a Down quark. The resulting particle then contains two Down quarks and one Up quark. This is a neutron.

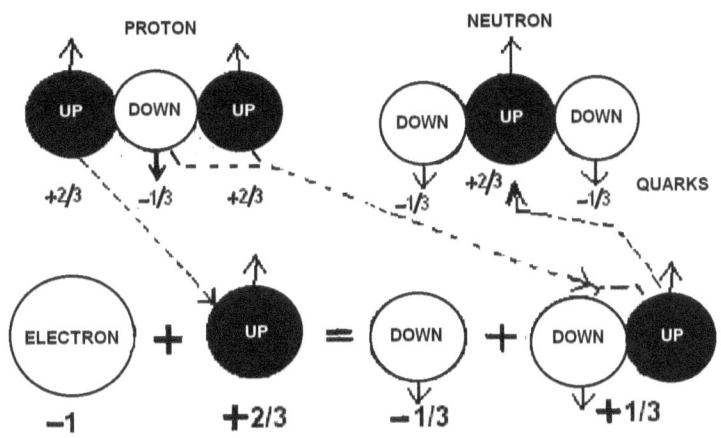

Figure 19

When a neutron combines with the nucleus of a hydrogen atom, it forms the isotope of hydrogen called deuterium, which is also a significant constituent of stars.

Figures 20 and 21 show two configurations of the deuterium nucleus. In Figure 20 only the electric fields would produce interactions with an electron. I do not think this would produce photons.

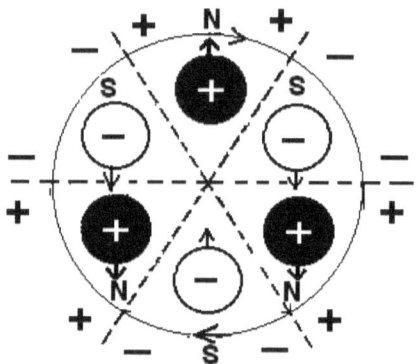

Figure 20

In Figure 21 below, the configuration would allow alternating interactions with an electron, from both electric and magnetic fields of the nucleus. The electron and the Up quarks must have the same magnetic field orientations.

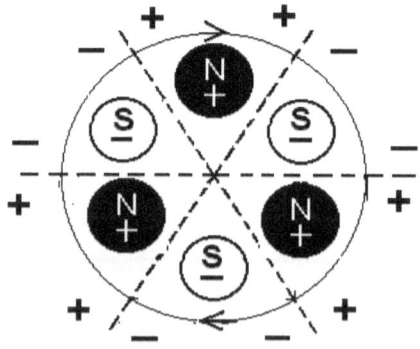

Figure 21

An approaching electron would experience six alternations of the fields for each rotation of the deuterium nucleus. The electron would have to rotate three times, during a single rotation of the nucleus to produce the same kind of interaction as is shown in Figure 17 for the hydrogen atom.

I assume that odd multiples of this spin rate of the electron, relative to the deuterium atom, such as nine and fifteen, would also produce similar effects.

I should repeat, that it is not the actual spin rate of the electron that is different. It is the relationship between the nucleus spin rate and that of the electron that changes. Since the deuterium nucleus has twice the mass of the proton, it is reasonable that it would rotate slower than the proton. Then the required electron distance and velocity, in the two atoms, should be similar when they produce photons.

These two atoms, hydrogen and deuterium, were the first ones that formed, after the start of our universe. They constituted the greatest majority of the atoms in the universe. However, the amount of hydrogen was far greater than that of deuterium.

These atoms formed gasses that gathered into large, globular concentrations. This began the process of star formation. The weight of a globular volume of gas produced a high pressure at the center. It was the presence of the deuterium at the center that allowed an easier start of nuclear fusion. Two deuterium atoms combined to form an atom of helium, which has a slightly smaller mass than two deuterium atoms.

The net loss of mass turned into energetic, subatomic particles, which heated the core of the star. The increased heat exerted enough pressure to resist the inward pressure of the great mass of the star.

There is a rare reaction, involving a proton combining with a tritium atom that can occur. Tritium is a combination of a proton with two neutrons. However, the most prevalent energy producing reaction in the core of a star is when two deuterium atoms combined forming helium.

The structure of the helium atom would have to be compatible with handling two valence electrons. The electrons would need to have opposite spin axis polarities. What I show in Figure 22 is a possibility. It is essentially two deuterium atoms, with one flipped over and sandwiched with the other. I assume that this nucleus rotates slower than the deuterium nuclei did before.

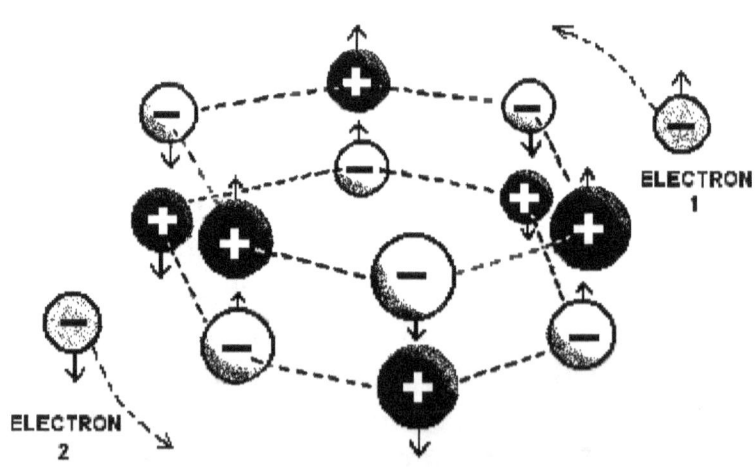

ELECTRON 1

ELECTRON 2

Figure 22

The helium nucleus rotates in the same direction as the two electrons. One of the electrons will have interacted with one of the deuterium rings and produced one or more photons. The other electron has the opposite spin polarity and will have interacted with the other deuterium ring to produce photons. Both electrons will be in similar orbits, relatively close to the nucleus. The magnetic fields are essential for the correct interactions to occur.

Over an extended period of time, the supply of deuterium became depleted and the core temperature started to decrease. The weight of the star would compress the core further until the next stage of nuclear fusion started. Helium atoms started to form heavier atoms, with the generation of enough additional heat to stop further compression.

I have also tried to extend this concept to the lithium atom. Since it contains an additional proton and electron, I assumed an additional neutron. Therefore, I added another deuterium ring to the helium atom, with the results shown in Figure 23.

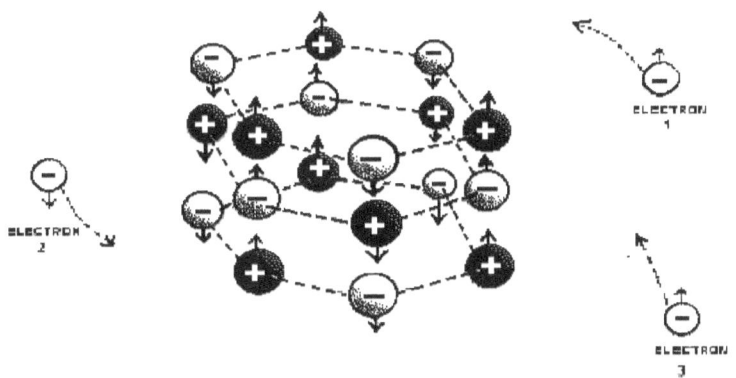

Figure 23

It seems that this configuration of lithium, with three protons and three neutrons, occurs a very small percentage of the time in nature.

The majority of lithium nuclei contain a fourth neutron. This extra neutron could be inside the nucleus, as shown in Figure 24. This should add to the stability of the nucleus.

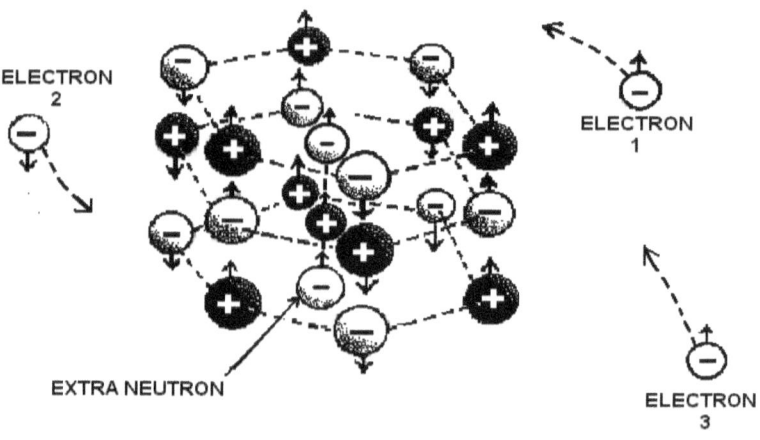

Figure 24

The first two electrons of the lithium atom are in the innermost orbital region, as in the helium atom. The third electron must occupy an orbit farther out, reacting to the electric and magnetic fields that are the result of the alternating vertically oriented protons and neutrons. The electron (number 3) will be attracted by the net positive charge of a proton but repelled by their similar magnetic polarities. It will also be repelled by the effective net negative charge of a neutron but attracted by their oppositely oriented magnetic polarities.

I have included tables of the elements at the end of this chapter. These tables give more details about the various atoms. See pages 121 to 123.

The next elements, after lithium, shown on the tables are beryllium and boron. Each adds a single proton and neutron to the nucleus of the previous atom. Of course, they also have one more valence electron than the previous atom.

The following element, carbon, adds only a proton to the number in the boron nucleus. This evens up the number of protons and neutrons to six of each. This is called, carbon twelve. Two other isotopes of carbon exist, having one or two additional neutrons.

The next two elements, nitrogen and oxygen, each add a combination of a single proton and neutron to the previous nucleus.

The heavier elements all have increasing numbers of neutrons, in addition to the number of protons. I feel that there might be a physical picture of how atoms are built.

Based on the above information, I have reached some conclusions and have made the following assumptions:

1. As the atomic nuclei get larger, with more protons and even more neutrons, they must build up a fixed structure that exposes all the protons on the exterior surface of the structure. This provides the mechanism for controlling the orbiting electrons in their proper locations.

2. The protons, with neutrons interposed, are assembled into variously sized rings. This configuration will produce the alternating fields to interact with the orbiting electrons.

3. The surplus neutrons will accumulate within the rings and could assemble into a support and connecting structure for the rings.

4. The pattern of the electron numbers in the orbital shells will have been caused by the pattern of the protons in the associated rings of each element.

5. The fact that the electrons in each of the allowed orbits are in pairs, suggests that they have oppositely directed, magnetic polarities. I have assumed that the ring structure will also have pairs of rings with oppositely directed, positive quark polarities.

6. The larger the proton rings of the nucleus become, the larger their orbital electrons' spin rates should become to cause photon emission.

To illustrate these ideas, I have adopted a simplified way of drawing the protons and neutrons in the various sized rings, as rectangles with dark dots for Up quarks and lighter dots for Down quarks. It should be realized that the rings probably assume a circular shape, with alternating positive and negatively charged quarks. In Figure 25 the simplest 2-proton and 3-proton rings are shown in both formats.

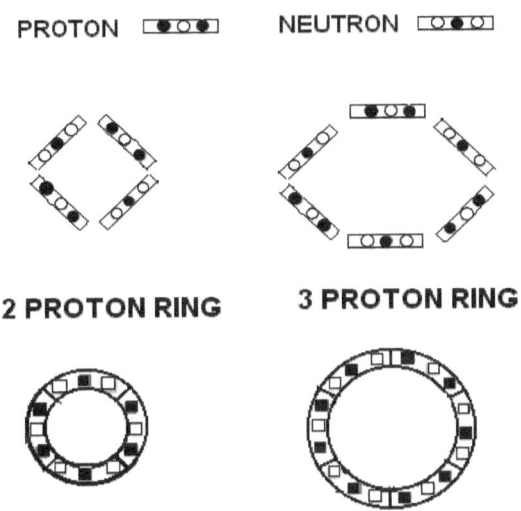

Figure 25

Similarly, in Figure 26, I show the next rings, with increasing numbers of protons.

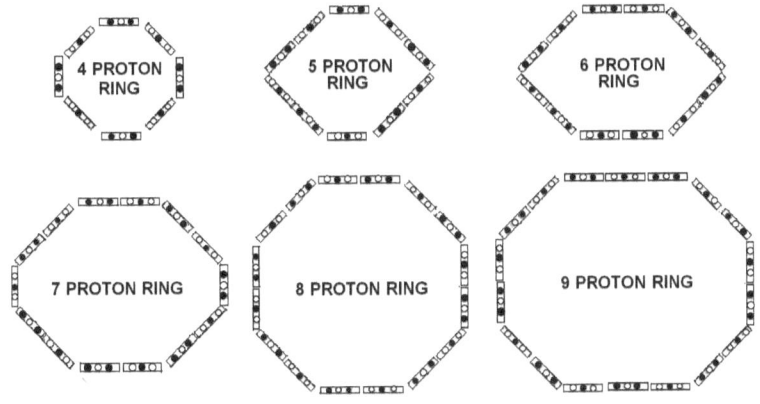

Figure 26

I am assuming that electrons in the closest orbits, controlled by the one proton rings, have a minimum of three times the spin rate of the nucleus when they emit photons. The larger rings should have electrons with spin rates that increase by a factor proportional to the number of protons in each ring. This should give these more distant electrons a stronger magnetic field to better interact with the nucleus and emit photons.

Illustrated in Figure 27, is an assembly of neutrons, having a linear and branching configuration. This could act as the central axis of the nucleus, with the branches connecting to the rings. The negatively charged Down quarks at the ends of the neutrons should be attracted to positively charged Up quarks.

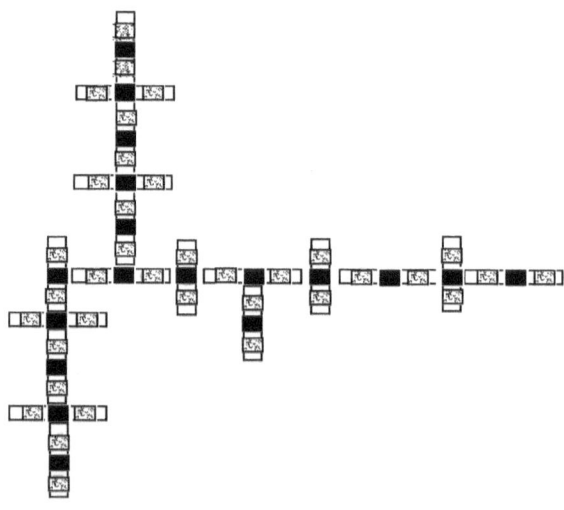

Figure 27

How a ring would fasten to a neutron branch is shown in Figure 28. The center neutron would be part of the axial neutron string.

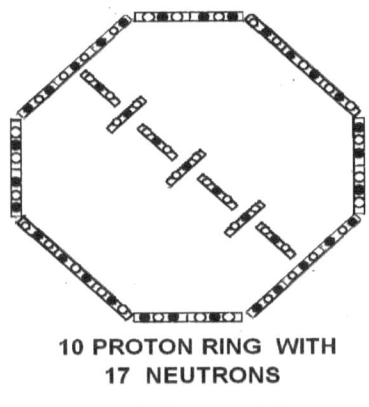

10 PROTON RING WITH
17 NEUTRONS

Figure 28

Pairs of rings with the same number of protons, but with opposed magnetic polarities, should hold together the same as the rings in the helium example. (See Figure 22.) But if there is too great a disparity in the proton number, from one ring to the next, some additional neutron bracing might occur, such as is shown below.

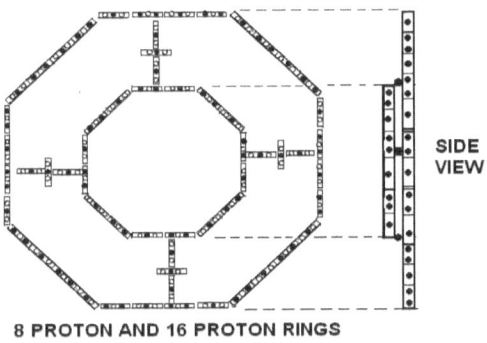

SIDE
VIEW

8 PROTON AND 16 PROTON RINGS

Figure 29

In Figure 29 a 16-proton ring is shown adjacent to an 8-proton ring. Four linear neutron strings connect the two rings together. It should be noted that the negative quarks at the ends of the neutron strings, connect to positive quarks at each end. I have included a side view of the rings, which I will discuss below.

Another simplifying technique used to help visualize the various atomic structures is to show the side view of a ring. It is shown as a vertical rectangle. The number of protons in each ring is given. This is shown in Figure 30.

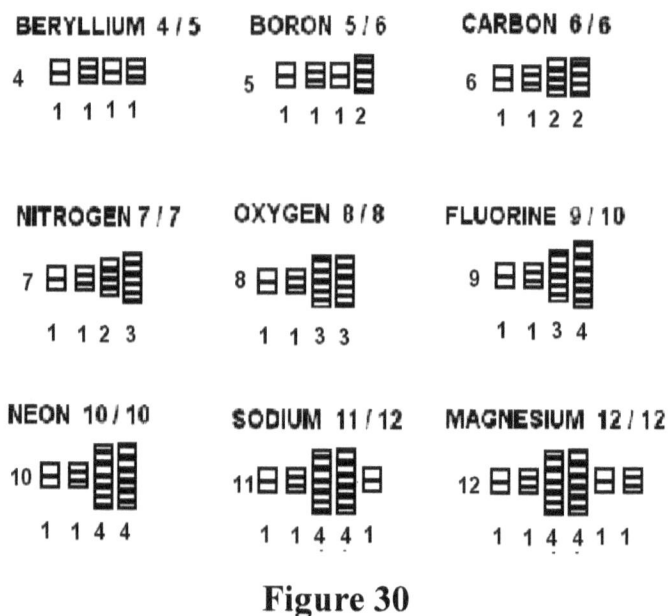

Figure 30

This figure starts with the beryllium nucleus in the upper left corner. This follows the 3-proton nucleus of lithium, as shown in Figure 23. But in Figure 30, the axis has been shift to horizontal instead of vertical. Beryllium is indicated to have 4 protons and 5 neutrons, and is shown with four 1-proton rings.

The neon nucleus is shown with two 1-proton rings and two 4-proton rings. These rings control first and second electron shells that are completely filled. Thereafter, the first proton rings of all the elements are like this. Since this is such an important interface, I show in Figure 31 how the one and four proton rings can be tied together.

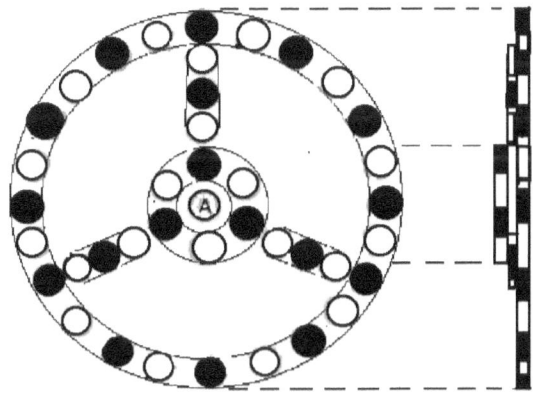

Figure 31

It can be seen that there are six quarks in the one-proton inner ring and twenty-four quarks are in the four-proton outer ring. Therefore, three neutrons are required to bridge between the corresponding pairs of positive quarks. An "A" indicates an axial quark in the center.

The next nucleus, in Figure 30 is sodium. It has a single proton ring added to the right end, controlling the valence electron in the highest orbit.

Magnesium has a second 1-proton ring added. The subsequent elements increase their number of protons in these rings until a total of eight is reached, with two, 4-proton rings. I assume that the following elements will increase the size of their proton rings according to the pattern of electron increases shown in Tables 1.

This sequence would repeat in the core of stars until the final fusions produced iron atoms. This was the beginning of the end for the star, since iron would not support the kind of fusion that produced energy. Many of these first generation stars would eventually explode and disperse the various elements into space where they would mix with the hydrogen clouds. From these clouds the second-generation stars were born.

In these stars the nuclear process in the core made neutrons that bombarded the heavier elements, such as iron. This produced even heavier elements over time.

For completeness, I am showing some of the heaviest nuclei in Figure 32. These are listed in Table 3.

Radon is the heaviest of the inert gasses. It shows the pattern of completed shells, from 2 to 8 to 18 to 32, without a final valence shell. Uranium is the heaviest, naturally occurring element. Plutonium is the familiar, man-made element.

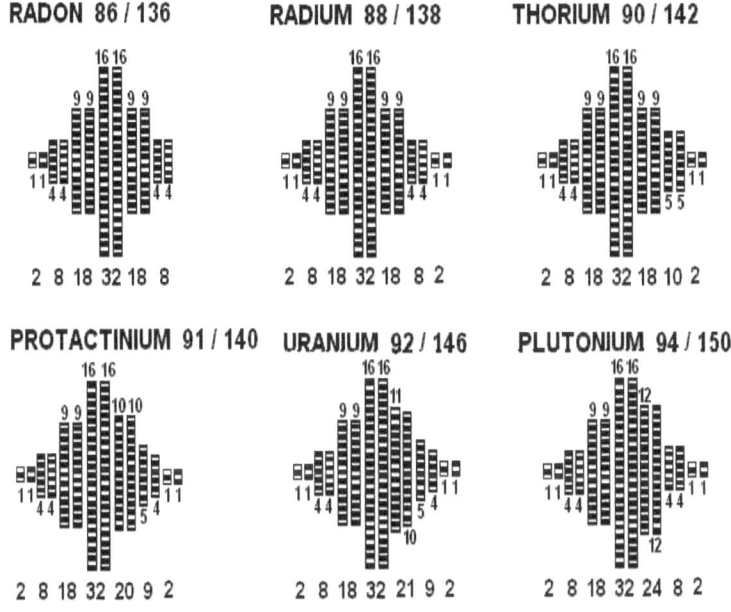

Figure 32

In all of these nuclei one can see how many additional neutrons are included, by noting the second number given. For example, in uranium, with 92 of the neutrons paired with the protons, the remaining 54 neutrons are inside the nucleus.

Note that in all of the above elements the maximum size rings occurs near the middle of the ring structure. They have 32-protons in a pair of 16-proton rings. The following rings get progressively smaller and control electrons in the higher orbits.

The electrons for the 16-proton ring should have a spin rate of about 48 times the nucleus spin rate when their photons are emitted. The electrons in the higher orbital shells should have spin rates at least this large.

A one-proton ring to control a valence electron in the highest orbit should have an electron with a spin rate much larger than the spin rates of the electrons in the lowest orbits. The greater spin of this electron should give it a larger magnetic field strength to interact with the nucleus.

For one last point of interest, I am including Figure 33. It shows the heaviest man-made nucleus that I have found. It is called unumbium and was made by smashing zinc and lead nuclei together. This is a very short-lived element with a half- life of about a quarter of a second.

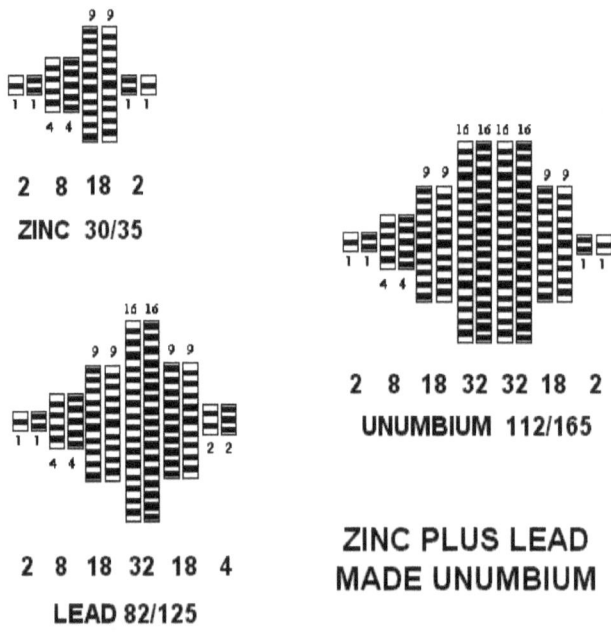

2 8 18 2
ZINC 30/35

2 8 18 32 18 4
LEAD 82/125

2 8 18 32 32 18 2
UNUMBIUM 112/165

ZINC PLUS LEAD
MADE UNUMBIUM

Figure 33

It is interesting to note that the first four ring pairs of the lead nucleus became the front of the unumbium nucleus without change. Also, the last two ring pairs of the zinc nucleus became the back end of the unumbium. All the other parts of the lead and zinc nuclei went into the additional pair of 16-proton rings of the unumbium.

This picture of atoms presents some questions to be considered, such as, "What makes the nucleus spin?" Remember that heavier atoms were built up, in the stars, by adding helium, deuterons and neutrons to lighter atoms. As the heavier, ionized atoms attracted electrons, some of the angular momentum, that the electrons gave up to go into orbit, was added to the angular momentum of the nuclei. This increased the spin of the nuclei.

Another question is, "How do the new electrons know where to go into orbit?" Modifying an atom, which already had its full complements of electrons, made a new ionized atom. An electron approaching the nucleus of this ionized atom, would find that only the ring with the added proton was lacking an electron. The new electron would avoid the other electrons and respond only to this ring, until a stable, synchronized, orbit was achieved.

All the electrons in an atom can be visualized as being in a flat halo, circling the nucleus in unison. Each electron would tend to stay in its own place, in a relatively stationary orbit over its own proton ring.

Any tendency of an electron to deviate from this stability would be corrected by its interaction with the multitude of electro-magnetic fields. The above concept gives a consistent, logical approach to understanding how atoms may be structured.

TABLE 1

	Elements	Atomic Nuclei		Isotope Neutrons	Electrons in Orbital Shells					
		Protons	Neutrons		No.1	No.2	No.3	No.4	No.5	no.6
H	hydrogen	1	0	1 and 2	1					
He	helium	2	2	1	2					
Li	lithium	3	4	3	2	1				
Be	berylium	4	5	6	2	2				
B	boron	5	6	5	2	3				
C	carbon	6	6	7 and 8	2	4				
N	nitrogen	7	7	8	2	5				
O	oxygen	8	8	9 and 10	2	6				
F	fluorine	9	10	9	2	7				
Ne	neon	10	10	11 and 12	2	8				
Na	sodium	11	12	11 and 13	2	8	1			
Mg	magnesium	12	12	13 and 14	2	8	2			
Al	aluminum	13	14	13	2	8	3			
Si	silicon	14	14	15 and 16	2	8	4			
P	phosphorus	15	16	15 and 17	2	8	5			
S	sulfur	16	16	17,18 and 20	2	8	6			
Cl	chlorine	17	18	20	2	8	7			
Ar	argon	18	22	18 and 20	2	8	8			
K	potassium	19	20	22	2	8	8	1		
Ca	calcium	20	20	22 to 28	2	8	8	2		
Sc	scandium	21	24	23 to 28	2	8	9	2		
Ti	titanium	22	26	24 to 28	2	8	10	2		
V	vanadium	23	28	25 to 29	2	8	11	2		
Cr	chromium	24	28	26, 29 and 30	2	8	12	2		
Mn	manganese	25	30	28	2	8	13	2		
Fe	iron	26	30	28, 31 and 32	2	8	14	2		
Co	cobalt	27	32	29 - 34	2	8	15	2		
Ni	nickel	28	31	28 - 37	2	8	16	2		
Cu	copper	29	35	32 - 37	2	8	17	2		
Zn	zinc	30	35	32-42	2	8	18	2		
Ga	gallium	31	39	35-41	2	8	18	3		
Ge	germanium	32	41	34-43	2	8	18	4		
As	arsenic	33	42	38 - 45	2	8	18	5		
Se	selenium	34	45	39 - 47	2	8	18	6		
Br	bromium	35	45	41-50	2	8	18	7		
Kr	krypton	36	48	41-54	2	8	18	8		
Rb	rubidium	37	48	44 - 53	2	8	18	8	1	
Sr	strontium	38	50	47 - 56	2	8	18	8	2	

TABLE 2

	Element	Atomic Nuclei			Electrons in Orbital Shells						
		Protons	Neutrons	Isotope Neutrons	No. 1	No. 2	No. 3	No. 4	No. 5	No. 6	No. 7
Y	yttrium	39	50	47-56	2	8	18	9	2		
Zr	zirconium	40	51	47-58	2	8	18	10	2		
Nb	niobium	41	52	49-56	2	8	18	12	1		
Mo	molybdenum	42	54	49 - 59	2	8	18	13	1		
Tc	technetium	43	55	52 - 57	2	8	18	14	1		
Ru	ruthenium	44	57	52 - 62	2	8	18	15	1		
Rh	rhodium	45	58	56 - 61	2	8	18	16	1		
Pd	palladium	46	60	56 - 64	2	8	18	18			
Ag	silver	47	61	57 - 64	2	8	18	18	1		
Cd	cadmium	48	64	59 - 70	2	8	18	18	2		
In	indium	49	66	63 - 68	2	8	18	18	3		
Sn	tin	50	69	62 - 76	2	8	18	18	4		
Sb	antimony	51	71	67 - 78	2	8	18	18	5		
Te	tellerium	52	76	67 - 82	2	8	18	18	6		
I	iodine	53	74	69 - 83	2	8	18	18	7		
Xe	xenon	54	77	68 - 84	2	8	18	18	8		
Cs	cesium	55	78	74 - 84	2	8	18	18	8	1	
Ba	barium	56	81		2	8	18	18	8	2	
La	lanthium	57	62		2	8	18	18	9	2	
Ce	cerium	58	82		2	8	18	20	8	2	
Pr	praseodymium	59	82		2	8	18	21	8	2	
Nd	neodymium	60	84		2	8	18	22	8	2	
Pm	promethium	61	84		2	8	18	23	8	2	
Sm	samarium	62	88		2	8	18	24	8	2	
Eu	europium	63	89		2	8	18	25	8	2	
Gd	gadolium	64	93	85 - 97	2	8	18	25	9	2	
Tb	terbium	65	94		2	8	18	27	8	2	
Dy	dysprosium	66	97		2	8	18	28	8	2	
Ho	holmium	67	98		2	8	18	29	8	2	
Er	erbium	68	99		2	8	18	30	8	2	
Tm	thulium	69	100		2	8	18	31	8	2	
Yb	ytterbium	70	103	98 - 106	2	8	18	32	8	2	
Lu	lutetium	71	104		2	8	18	32	9	2	
Hf	hafnium	72	107		2	8	18	32	10	2	
Ta	tantalium	73	108		2	8	18	32	11	2	
W	tungsten	74	110		2	8	18	32	12	2	
Re	rhenium	75	111		2	8	18	32	13	2	
Os	osmium	76	114		2	8	18	32	14	2	
Ir	itidium	77	115		2	8	18	32	15	2	
Pt	platinum	78	117		2	8	18	32	17	1	
Au	gold	79	118	115 - 120	2	8	18	32	18	1	
Hg	mercury	80	121		2	8	18	32	18	2	

TABLE 3.

	Elements	Atomic Nuclei		Isotope Neutrons	Electrons in Orbital Shells						
		Protons	Neutrons		No. 1	No. 2	No. 3	No. 4	No. 5	No. 6	No. 7
Tl	thallium	81	123		2	8	18	32	18	3	
Pb	lead	82	125		2	8	18	32	18	4	
Bi	bismuth	83	125		2	8	18	32	18	5	
Po	polonium	84	125		2	8	18	32	18	6	
At	astatine	85	125		2	8	18	32	18	7	
Rn	radon	86	136	130 -135	2	8	18	32	18	8	
Fr	francium	87	136		2	8	18	32	18	8	1
Ra	radium	88	138	134 - 139	2	8	18	32	18	8	2
Ac	actinium	89	138		2	8	18	32	18	9	2
Th	thorium	90	142		2	8	18	32	18	10	2
Pa	protactinium	91	140		2	8	18	32	20	9	2
U	uranium	92	146	138 - 148	2	8	18	32	21	9	2
Np	neptunium	93	144		2	8	18	32	23	8	2
Pu	plutonium	94	150		2	8	18	32	24	8	2
Am	americium	95	148		2	8	18	32	25	8	2
Cm	curium	96	151		2	8	18	32	25	9	2
Bk	berkelium	97	150		2	8	18	32	26	9	2
Cf	californium	98	153		2	8	18	32	28	8	2
				Half Life							
Es	einsteinium	99	153	1 yr.	2	8	18	32	29	8	2
Fm	fermium	100	157	3 min.	2	8	18	32	30	8	2
Md	mendelevium	101	157	50 sec.	2	8	18	32	31	8	2
No	nobelium	102	157	2 min.	2	8	18	32	32	8	2
Lr	lawrencium	103	159	13 sec.	2	8	18	32	32	9	2
Rf	rutherfordium	104	157	5 sec.	2	8	18	32	32	10	2
Db	dubnium	105	157	4 sec.	2	8	18	32	32	11	2
Sg	seaborgium	106	157	.3 sec.	2	8	18	32	32	12	2
Bh	bohrium	107	157	.1 sec.	2	8	18	32	32	13	2
Hs	hassium	108	157	2 ms.	2	8	18	32	32	14	2
Mt	meiterium	109	157	3 ms.	2	8	18	32	32	15	2
Uun	ununnilium	110	159	10 ms	2	8	18	32	32	17	1
Uuu	unununium	111	161	____	2	8	18	32	32	18	1
Uub	unumbium	112	165	.3sec.	2	8	18	32	32	18	2

BLACK HOLES

Our Sun is a second-generation star. It is one of the smaller sized stars. It is used to compare the size of other stars. We call our Sun "one solar mass". The smallest stars are about one half a solar mass and the largest are more than one hundred solar masses in size.

The amount of nuclear fusion needed to withstand the weight of a star varies with the size of the star. Our Sun has enough fuel to burn for billions of years. The largest stars may have to use up all their fuel in less than a million years. When a star has used up its fuel, a number of different things can happen. Some stars become large cool red stars. Some become small hot dwarf stars. Stars that are in the range of ten solar masses can explode and then collapse, becoming a neutron star.

The most spectacular event occurs when a star of about one hundred solar masses explodes as a super nova. It may then collapse into a black hole. Mathematicians have done much of the theoretical work on black holes. Their equations for the characteristics of matter at the end of the collapse of a star into a black hole, predict that the size of the body goes to zero and the density becomes infinite. However, these bodies are said to have a finite mass. This is called a singularity. I think this concept is hard to believe. I prefer one that is simpler and more logical.

The nature of neutron stars may be the best place to start in understanding my idea about black holes. Neutron stars are thought to have a surface shell of iron atoms, which are the remnants of the depleted nuclear core.

Inside of the surface shell are the neutrons, essentially in contact with one another. In the collapse of the core the electrons of the heavier atoms are forced into the nuclei. The electrons and protons combine, forming neutrons.

It is theorized that at the center of the neutron star some exotic particles could be forming. I wondered what these exotic particles could be. This led to the idea of neutral particles, made up the next heavier quarks, becoming the predominant component at the center of the individual black hole from a collapsed star. This particle should be a hundred times more massive than a neutron.

The black hole should still have a layered structure with an outside layer of iron. Within that there should be a layer of neutrons and inside of that there should be the particles made up of the charm and strange quarks. I really need a shorter name for those particles, such as "charstrons", perhaps.

Neutron stars are said to result from the collapse of stars of about eight solar masses. After they collapsed, they had masses of a little over one solar mass. I am assuming that this was essentially the mass of their nuclear core. The remainder of their original mass must have been mostly hydrogen gas blown off in the super nova. I understand that these bodies ended up with a diameter of about ten miles.

A scientist named Schwartzchild derived an expression that gave the relationship between a body's mass and the distance from which light could no longer escape.

For a one solar mass neutron star with a diameter of about ten miles, the Schwartzchild limit is a four-mile diameter sphere. This would be within the body of the neutron star. Therefore, the neutron star should still be visible.

In the following figure the relative size of different mass stars is shown. Their masses increase with their volumes, and their volumes increase as a function of their diameters cubed.

STAR SOLAR MASSES

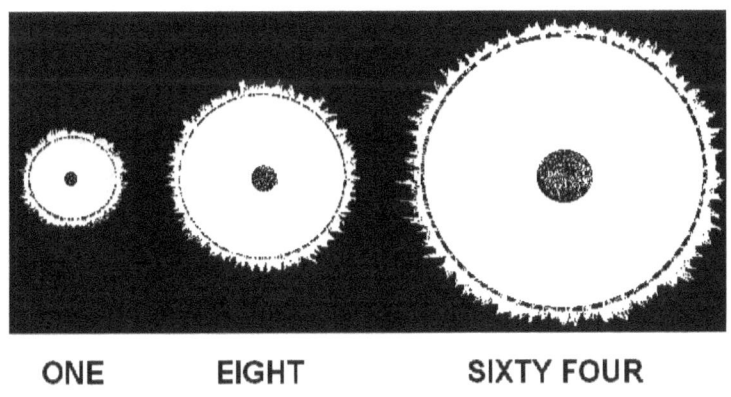

ONE   EIGHT   SIXTY FOUR

Figure 34

Black holes are supposed to occur some of the time when larger stars, in the range of fifty to one hundred solar masses, collapse in a super nova. I am assuming that a sixty-four solar mass star would have a core of about eight solar masses. If it had collapsed into a neutron star, which it might in a transient phase, it would have had a diameter twice that of the above neutron star. That is about a twenty-mile diameter. It should still have about the same mass of eight solar masses.

For this mass the Schwartzchild limit would be a sphere with about a thirty-two mile diameter. Therefore if this neutron star remained a neutron star it would be close to being a black hole.

However, assume that many of the neutrons in the center of this body were converted to charstrons. If the ratio, of neutrons being converted to charstrons, were about a hundred to one, the twenty-mile diameter of the eight solar mass body, would be reduced by a factor of about four. That is assuming the individual charstrons were about the same size as neutrons. It would stop collapsing at about a diameter of five miles. This would definitely be a black hole.

In the above examples, I have assumed that the Schwartzchild equation is valid. Since it is partially based on an Einstein theory it may not be.

If this first kind of black hole were in a situation where it was absorbing more and more matter from adjacent stars, it might reach a state where it would collapse again. A large number of "charm" and "strange" quarks would be converted into a smaller number of the more massive "top" and "bottom" quarks. These quarks would become neutral particles about one hundred times as massive as the charstron particles. Perhaps, we could call them "tobotrons". They might reduce the diameter of a charstron black body by a factor of four in becoming an equivalent tobotron black body.

I am assuming that it would be configured with a layer of iron on the outside. Within that there should be a layers of neutrons and charstrons.. The tobotrons would form within the layer of charstrons.

Perhaps a different name, other than black holes, would be appropriate. Astronomical bodies are commonly named using the suffix, "oid". That is like asteroid, meteoroid and planetoid. I would like to suggest that neutron stars be called "neutroids", charstron black holes be called "charstroids" and tobotron black holes be called "tobotroids".

The particles that would result from the conversion of quarks of one type to those of another are shown in the following figure. There is the possibility that the charstron quark charges are 100 times those of the individual neutron quarks. Also, this would imply that the charges of the tobotron quarks could be 100 times that of the charstron quarks and 10,000 times that of the neutron quarks.

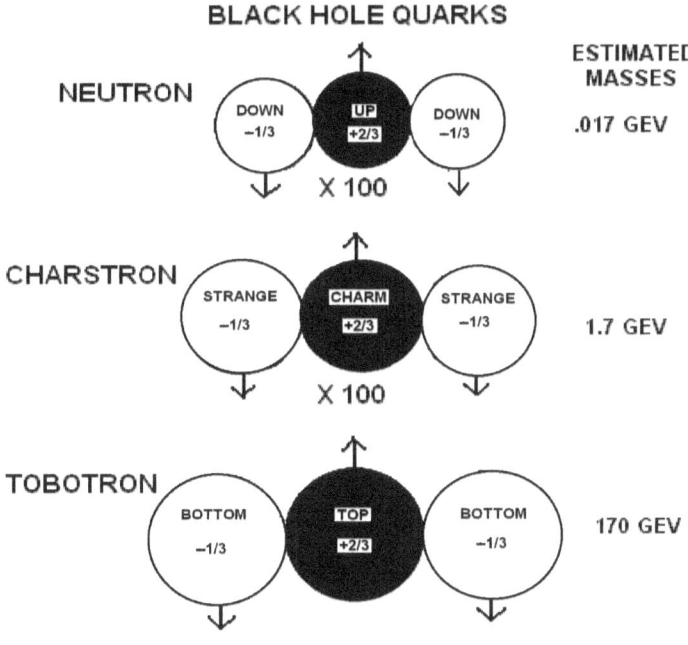

Figure 35

It is interesting to note that the above tobotron has about the same energy equivalent mass as that attributed to the Higgs particle, 170 GEV.

There is evidence that there are massive black holes at the center of most galaxies. They appear to have masses in the order of millions of times that of our Sun. This suggests to me that these could be black holes of the second kind (tobotroids). If a hypothetical tobotroid of ten solar masses would have a diameter of about one mile, a ten-million solar mass tobotroid, in the center of the galaxy, would have a diameter of about one hundred miles.

Assume that there are a hundred billion solar masses of stars in the galaxy. If all were ingested into this central black hole it would have a diameter increased by a factor of the cube root of 10 thousand, which is about 22. The final diameter of the galaxy weight black body would be about 2200 miles. (We could call this a galaxoid.)

If a hundred billion such galaxoids combined into a single massive black hole, its diameter would have increased by a factor of the cube root of a hundred billion (about 4500). Therefore the final diameter of a universe weight black body (univoid?) would be about ten million miles. This supposedly accounts for the present visible universe. Since most of the gravitons will remain dark, assume that they represent 7/8 of the total volume. This suggests a final diameter of twenty million miles.

It is conceivable that even more massive quarks could form at the center of such a body. These could create another type of neutral particles out of tobotrons. This could reduce the diameter of the universal black body even further.

Consider how black holes will interact with other matter. Gravitons will impact on a black hole from all directions. The gravitons will enter through the spaces between the surface atoms and make innumerable elastic collisions within, transmitting their momentum into the black hole.

Any material body in the vicinity would be accelerated toward the black hole, since no gravitons would be coming out of the black hole to impede them. The approaching bodies would achieve exceedingly high velocities and be assimilated into the black hole upon impact.

When black holes are in the same vicinity, they will tend to shield each other from gravitons coming from the directions beyond them. These gravitons will impart momentum to the black holes they fall upon and cause the black holes to move towards each other, until they merge.

It is evident that black holes would accumulate gravitons. These gravitons will have come from their own universe, from adjacent universes, and even from previous universes that no longer exist. The gravitons will be recycled, liberated when the black holes containing them turn into a Big Bang.

BIG BANG

We can look at the big bang in a new way. Instead of assuming a collapse of the universe size black hole, assume that something causes it to open up. Perhaps the impact of a large incoming mass (another black hole) could cause the black holes to crack open.

Another possibility is that if the flux of incoming gravitons dropped below a certain level, the internal pressure could crack the surface layers. This would release the charstrons and the tobotrons from containment. They would erupt in all directions.

The individual tobotrons could become a large number of charstrons. The charstrons could become many neutrons. These would race off in every direction at a velocity, greater than the present speed of light. This expanding sphere of high velocity neutrons would have enumerable collisions occurring.

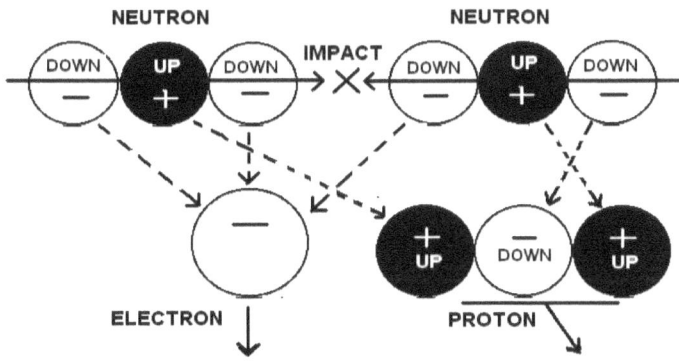

Figure 36

Figure 36 shows how a collision between two neutrons could produce a proton and an electron. Later, neutron collisions with protons could produce deuterons.

Eventually, electrons will combine with protons and become hydrogen. Also, electrons and deuterons will form deuterium. Star formation could then begin.

A soup of super-luminal neutrons could be filling the universe to this day. They can be entering all objects from all directions. They could make occasional impacts on the atoms of matter imparting momentum. These particles could be what I have called gravitons.

When I described how black holes accumulated gravitons, it could have been neutrons being accumulated. This would have increased the conversion of neutrons to charstrons and eventually increasing the number of tobotrons.

At the instant of the big bang, if all of the particles in the twenty million mile diameter black body turned instantly into neutrons, essentially in contact, the volume would have expanded by a factor of ten thousand. The diameter would be almost five hundred million miles.

After a relatively short period of this high concentration of neutrons, the density would rapidly decrease. At the present time the contribution of the big bang gravitons may only be a small portion of the total number of gravitons. Probably, the gravitons from the surrounding universes make up the rest.

GRAVITONS

At the instant of the big bang, the first force that is supposed to have emerged is gravity. It has been assumed that this was a force of attraction. But by my assumption, it was a repellant force and would initiate rapid expansion.

There has been an attempt to replace the "Inflationary Theory" in the Big Bang. One of the ideas is that at the very beginning of the universe, the velocity of light was far higher than it is now. With my concept of the speed of light being controlled by the speed of gravitons, this would imply that the speed of gravitons, at that time, was much higher than it is presently. It may be that over a long period of time the multiple collisions made by the gravitons would reduce their momentum slightly. This would cause a reduction in their velocities.

With further cooling, (reduced velocity impacts), the other forces would emerge, and then the primary particles would form. These particles would interact with the gravitons. Most of the gravitons would be scattered by the particles and remain among them.

As the universe evolved, the greatest portion of the visible mass of the universe would be in the central part of the distribution, at some average radius. Within this region, bodies would be impacted by gravitons coming from every direction. Only in the innermost and the outermost parts of the population would there be an imbalance of the graviton impacts.

Bodies in the innermost region would tend to be retarded and their rate of expansion would be slowed by the predominant direction of the graviton flux being from the central region.

The bodies that are in the farthest out region will also be affected by the graviton flux in an unbalanced way. They will be accelerated by the gravitons coming from the central region behind them.

What gravitons are is a good question. They would have to perfuse our space but not have been detected, except for the effect of gravitation they produce. Originally, I thought that the theoretical particle, the Higgs Boson was the graviton. If a large number of tobotrons had survived the big bang, they could be an equivalent of the Higgs particle and serve as gravitons.

The vastly more numerous neutrons that were released in the big bang are much better candidates to be gravitons. I have assumed that gravitons are neutrons traveling with velocities greater than the speed of photons.

What could be the interaction of a graviton with a photon? What could the interaction of the graviton with an atom of matter be like?

The graviton (neutron) has two negative, Down quarks linked by a positive Up quark. The Down quarks are in a quasi-stable configuration and will tend to oscillate and then cause the graviton to spin.

The following figure shows how the electro-magnetic fields surrounding the spinning graviton might interact with photons while passing them.

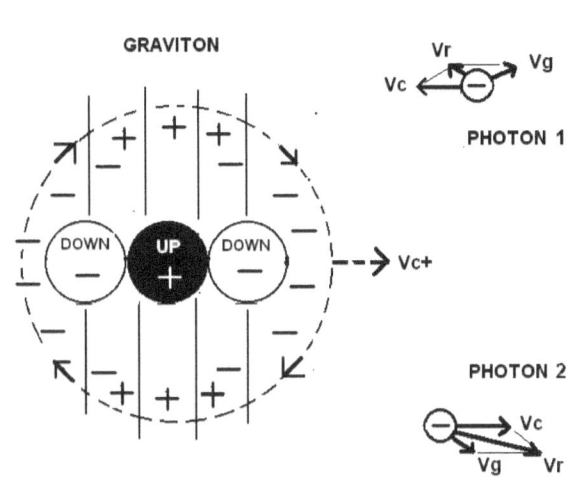

Figure 37

The rotating electric fields of the graviton would be predominately from the negative quarks. The graviton is moving to the right with the velocity Vc+. The upper photon is moving to the left with a velocity Vc. The vector Vg shows the effect of the electric field of the graviton upon the negatively charged photon. The magnitude and direction of the resultant velocity of the photon is shown as Vr. The photon has been slowed down and deflected away from the graviton.

At a different time than that of the above encounter, the lower photon is traveling with the velocity Vc in the same direction as the graviton. As the graviton approaches the photon the effect of the electric field of the graviton is to accelerate the photon and deflect it away.

Since the mass of the graviton is so much greater than that of the photon, the velocity and direction of the graviton is not affected.

Because the graviton passes the lower photon with less of a relative velocity, it has more time to accelerate this photon. There is less time to decelerate the upper photon in their shorter duration encounter.

When a graviton approaches an atom of some heavier element the interaction is different. Occasionally a graviton may impact a nucleus. A graviton is more likely to be deflected by the negative charges of the orbiting electrons of the atom. The momentum of the graviton is so high that only the small amount of its momentum that is transferred to the atom will move the more massive atom a small amount, in this elastic collision. The graviton may not have to come very close to an atom to transfer some momentum to it.

It is now believed that about three quarters of the mass of the universe is dark. Dust clouds, black holes or neutrinos cannot explain this mass. It is also thought that something called dark energy exists which acts like a negative force of gravity. These ideas have been theorized because of observations of unexpected occurrences in distant super novas and galaxies.

My theory, that swarms of "faster than light" neutrons fill the universe and provide the effects of negative gravity, explains this. The graviton theory is consistent with the idea of dark matter. Neutrons do not absorb or emit light. They probably were the main product of the big bang.

However, the most likely source of most of the gravitons in our universe could be from outside our universe. Most of the mass in the universe could be gravitons.

A galaxy will tend to act as a sink for gravitons in that its black holes will absorb the gravitons. More gravitons will move towards a galaxy than will be moving away from it. This will cause a gradient in graviton flux density, which will cause the observed motions.

Starlight from more distant galaxies will be deflected by the graviton flux gradient around a galaxy. The path of light will curve from the denser graviton flux toward the less dense flux. This has been called a gravitational lens.

If one wanted to travel from Earth to some other star, it should be easier to reach one that is closer to the center of our galaxy, than one in the opposite direction. Our galaxy's graviton flux gradient should favor this.

MOTION

What is the effect of gravitons upon a body in motion? Assume that a small body is alone in space, and that it is in an unknown graviton flux. The body could be in a non-uniform graviton flux. There could be more gravitons coming from one direction than coming from another. Also, The velocities of the gravitons coming from one direction could be different than those coming from another.

The body would be accelerated by the gravitons, toward the source of the lower graviton flux or toward the source of the lower graviton velocities.

If the body is moving with a constant velocity, relative to some distant frame of reference, it is in a non-uniform graviton flux that compensates for the velocity of the body.

As long as the gravitons coming from any direction impart the same momentum to the body, as those coming from the opposite direction, there will be no change in the body's momentum.

This suggests that one of Newton`s laws of motion is only approximately correct. As long as the body has a velocity that is small compared to the velocity of gravitons, it will appear to remain in motion without change.

With a source of thrust, the body can accelerate and achieve a higher velocity. This will give the body an increased velocity relative to the gravitons that impact the front of the body. They will transfer more momentum to the body than the gravitons impacting the rear. This imbalance will provide an opposing force that the thrust must overcome.

If the thrust is discontinued, the body will decelerate until it reaches a velocity, where it is balanced by the graviton flux differential.

The momentum of a neutron, with the velocity C, impacting the front of a body that is traveling at a velocity V, is the mass of a neutron (Mn) times (C+V). The momentum of a neutron impacting the rear of the body depends on (C-V). The net momentum, causing deceleration, is the difference between these, that is:

$[(C+V) - (C-V)]$ (Mn) $= 2V$ (Mn).

If sufficient thrust were applied for the body to reach a velocity that approached the speed of light, the momentum of the gravitons impacting the front would become twice as great and the momentum of those impacting from the rear would approach zero. It would take a very large thrust to maintain this velocity.

The required force for the body to continue accelerating, beyond the speed of light, will depend more on the increase in the momentum of the body than on the impacts of gravitons against it. Therefore with sufficient thrust a vehicle should be able to attain faster than light travel.

This is in direct contradiction to Einstein's theory of Special Relativity which states that an infinite amount of energy is required to accelerate a body to the speed of light. Because of this, it is important to either verify or refute my new theory of gravity and motion, for the benefit of man's future progress in space.

FASTER-THAN-LIGHT

This is a highly speculative idea that follows up on my suggestion that faster than light (FTL) travel may be possible. As I indicated in the chapter on motion, in order to maintain a given velocity, a vehicle will have to provide a constant thrust of the proper magnitude. To do this for inter-stellar travel, the vehicle must have a continuous supply of fuel. My idea is to tap the gravitons of space to provide this fuel.

I hope that a method can be developed that will cause, under controlled conditions, the kind of collisions between FTL neutrons that is described in the chapter on the Big Bang. (See Figure 36.) The protons and electrons produced will be the inputs to proton accelerators. The following diagram shows a possible configuration of a FTL propulsion system.

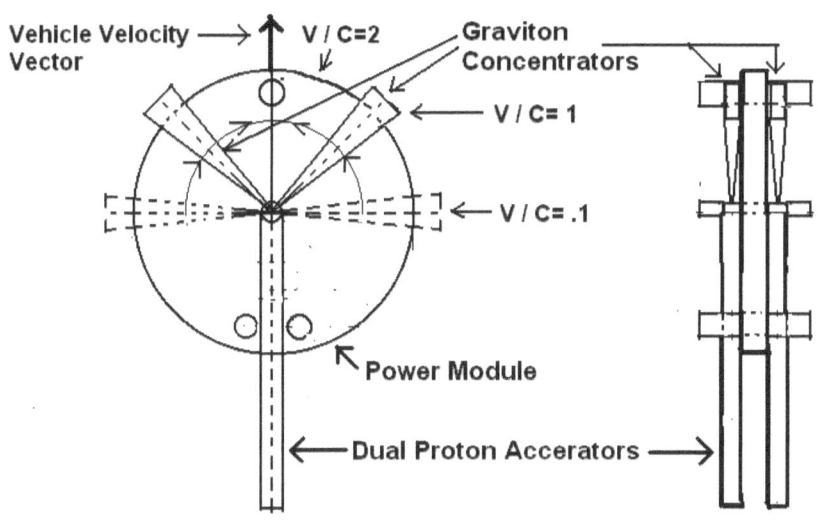

Faster Than Light Propulsion System

Figure 38

The main engineering challenge will be to concentrate the neutrons going into the collisions so as to provide adequate output protons.

As indicated, the vehicle is moving upward with a vertical velocity, V. The graviton concentrators are shown in a position appropriate for a V / C ratio of 1.

The horizontal components of the graviton flux will enter the concentrators from the right and the left. With the ship`s velocity equal to the graviton velocity, the gravitons will follow the 45 degree path to the apex of the concentrators. I will call them horns hereafter.

With the ship starting out at a low velocity the horns will be in the position shown for V / C = .1 ratio. With increased velocity the horns will be moved forward as required. Eventually, a vehicle velocity of twice the speed of light may be reached with the horns in the position indicated as V / C = 2. The horns may be oriented separately but will usually be placed symmetrically. They will be capable of moving through arcs of almost 180 degrees.

A second pair of horns and another proton accelerator has been included in this configuration. This doubles the propulsion with only a little additional ship mass.

The ability of a horn to concentrate all the gravitons entering the aperture into a narrow beam at the collision point is critical. I can think of several possible approaches to accomplish this. The horns might have a passive design that uses their shape and material characteristics to deflect the neutrons toward the vertex. It is more likely that an active approach will be needed.

This will probably involve various electric and magnetic fields. These might be steady state fields or require higher energy pulses.

The region of the graviton collisions will also require an innovated design. The area will have to be shielded. The protons will have to be contained and directed down the accelerators. The electrons will have to be separated out and directed into electro static fields or into power regeneration.

The above describes how the vehicle accelerates to FTL speed. For the ship to decelerate, reducing or discontinuing the thrust will cause a reduction in velocity due to the head-on impacts of the graviton flux.

After the vehicle has reached a relatively low velocity, if a more rapid deceleration is needed, the ship may be rotated so that the proton accelerators face the forward direction. Their thrust will then cause a more rapid deceleration. To provide the fuel for this activity the horns can be moved to the appropriate position near the proton accelerator end of the ship.

The various problems of life support, recycling water and regenerating oxygen have been worked upon already. I assume that the primary long-term power source will be nuclear. For this to be a feasible inter-stellar vehicle, a great amount of back up capacity and redundancy will be required. Trips of many years, even at FTL speeds, will be required to reach the closest star systems.

In the following diagram I show how combining duplicate versions of the basic vehicle propulsion will provide a degree of extra safety.

CREW QUARTERS

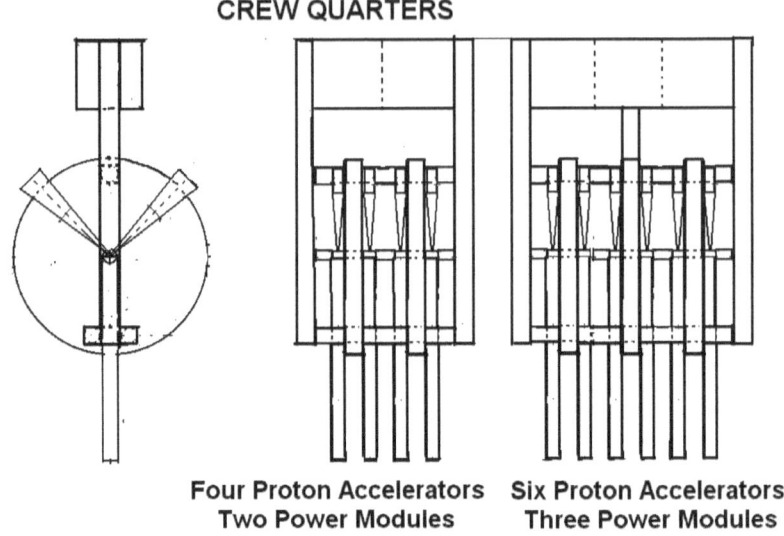

Four Proton Accelerators   Six Proton Accelerators
Two Power Modules   Three Power Modules

MODULAR FASTER THAN LIGHT VEHICLES

Figure 39

As shown, the habitable areas have been placed forward, away from the internal sources of radiation. The quarters for the crew may be separated into several compartments, each with control of one or more of the propulsion units.

If the technique of flipping the vehicle over, to be able to use thrust for deceleration, is done at very high velocities, the floors would become ceilings and the ceilings would become floors. If this is a requirement, the compartments could be designed to rotate 180 degrees around a horizontal axis, as this takes place.

The effects on the crew, of the acceleration and constant velocity of the vehicle and the probable forces relative to the Earth's force of gravity, are a concern.

The acceleration should be maintained at a tolerable level. At some high velocity the difference in graviton flux between the head-on and following gravitons, and its effect on the momentum of the vehicle and its occupants, should be equivalent to the force of gravity experienced on Earth. If this occurs at a velocity close to the speed of light, then the crew might tolerate the additional force as the velocity is increased further. At the speed of light the vehicle and its occupants will experience a maximum retarding force, since only head-on graviton impacts will occur.

After the speed of light is reached, increasing the velocity will increase the momentum of the ship and crew but not the momentum of the head-on gravitons. The effective retarding force on the ship and crew may actually decrease with increased velocity.

The increased momentum of a body in the vertical direction may restrict most motion to the horizontal plane. This would also apply to photons and electrons in electronic control systems.

If approaching the speed of light produces intolerable effects, the maximum velocity will have to be limited to some lower value. Hopefully, this would be at some appreciable fraction of the speed of light.

CONCLUSIONS

I have presented my ideas of a worldview that may not be as mathematically elegant as those of Albert Einstein, but hopefully they will be more readily visualized. I see things differently then those who use the Relativity equations unquestionably as the starting point for their extrapolations. My ideas may not all prove to be correct, but they have the beauty of simplicity and being understandable.

Einstein was able to develop his Special Relativity in less than a year. It took him ten years to develop General Relativity, to tie gravity in with his first approach. He then spent the rest of his life trying, unsuccessfully, to truly unify gravity with the other forces of nature. I am the most uncomfortable about his use of time as an independent variable in equations that predict hypothetical effects of velocity on time and the mass of objects.

In summarizing the ideas presented here:
I believe that the three dimensional view of objects moving out into a universe of non-expanding space is useful to overcome the dead end of the expanding space model, which predicts that every other body in space will eventually disappear from our view, moving away at a speed faster than light.

I believe that our universe is a naturally occurring process that must have a means of reproducing itself. That is why the idea of multiple universes surrounding ours is a reasonable possibility.

The idea that photons have mass is not really too new. It is only the idea that they are limited to a velocity determined by gravitons that is new.

I consider the idea of gravitons, acting as a repellant force that imparts momentum to objects and produces the action of gravity, is the most significant of the ideas presented here.

The concept of black holes being knowable as objects that are produced by a natural extension of known processes is a reasonable one. This ties together a progression; from a large star mass black hole, to a galaxy mass black hole, to an universe mass black hole, and to the Big Bang.

I conclude that gravitons are neutrons traveling with velocities greater than the speed of light. These gravitons could be the mysterious dark mass that has been theorized as constituting the vast majority of the mass of the universe.

I suggest using electro-magnetic coupling, instead of "gluons" to provide a useful configuration of quarks, within protons and neutrons. This led to my idea of how the interaction of a proton with an electron produces photons in a hydrogen atom and also that the photons are very small bits of the electrons emitting them. I also, derived a concept of how electrons and nuclei interact and establish the electron orbits in atoms. I describe a ring structure for the nuclei of all the elements that can account for the structure of the electron orbits.

Finally, in a detailed explanation of how gravitons might interact with matter, I present a new theory of motion that modifies Newton's law of motion. My theory also appears to indicate that faster than light travel may be possible.

This book presents what may be the long sought, unified theory of gravity and electro-magnetism.

(This is from a self-portrait by the author)

Lenard Metzger was born and raised in Rochester, New York. He has lived there ever since, except for the years spent in service and college. He received a degree in physics from Ohio State University.

He was a long time employee of Eastman Kodak Company. After retiring he returned to his early interests in art, science and writing.

ADDENDUM B

Addendum B. contains the portions of "Common Sense Cosmology" that were <u>not</u> included in "Beyond Einstein".

Common Sense Cosmology

By

Lenard Metzger

Elemental Publishing
Rochester, New York

COMMON SENSE COSMOLOGY

Elemental Publishing
80 Westerloe Avenue
Rochester, NY 14620
(585) 473-9303

Library of Congress
Control Number: 2010930766

ID: 332616
lulu.com

ISBN: 978-0-557-51848-7

PREFACE

I have been trying to decide how to present the ideas that I have been thinking about for a long time. What I have decided to do is to use a narrative style rather than a textbook style. There is the question of what to call these ideas. They certainly do not have the status of "theories". "Hypothesis" is much too fancy a word. I am not sure about "conjectures". I will call these concepts or ideas and leave off the term "crackpot".

At about my ninth or tenth birthday, my sister gave me a book with a title something like "The Wonders of Science". I learned many wonderful things reading it. All the strange things people have believed, over the years, about the world and the universe, puzzled me.

How could people believe that the world was flat and that if you sailed a ship to the edge your ship would fall off? Why didn't they realize that all the water in the seas would pour off the edges and the ship would have nothing to sail on?

After it was generally accepted that the world was round, there was this idea that the Sun and the stars all revolved around the Earth. And even when it was finally understood that the Earth revolved around the Sun, the belief was that the stars were all on the surface of a crystalline sphere that rotated around us.

The thought then was that the region beyond this was heaven, the abode of the angels. Then people realized that some of the "stars" moved erratically, along different paths then those of the rest of the stars.

These objects were named "planets" and were thought to be on separate moving crystalline spheres.

It was only later that true astronomers described the real situation, that the Earth is one of a group of planets circling the Sun and that the stars are at a very much greater distance away.

How could people have believed in crystalline spheres with nothing beyond them? In my teens I tried to visualize this but every time I reached a mental barrier in space I continued outward from the other side of the barrier. I soon concluded that there was no end to space.

When I first learned about what was thought to be the structure of atoms in matter, I was struck by how similar their structure was to that of our solar system, with the electrons circling the atomic nucleus, just as the planets circle the Sun. When I learned of galaxies and molecules, I thought how their relationships were similar to that of solar systems and atoms. Later, I thought, what if galaxies are organized similar to the way molecules are assembled into crystals? The final leap was to imagine the entire universe as a small object in some much larger scheme of things. I painted a picture of this concept and later wrote a poem about it.

Going back to the beginning, when I was around seven or eight years old, I was fascinated by the ham radio that my older brother had built. He got tired of my playing with it and unplugging the coils and putting them in the wrong sockets. So he bought a crystal set for me and helped me build my own little radio.

Later I looked at his copies of "Popular Mechanics" and "Radio and Electronics". While in high school, I started reading science fiction magazines, such as "Fantastic" and "Amazing". I progressed to more scientific kinds, such as "Astounding" and later "Analog". At that time I started reading "Scientific American" and have persisted with it to this day.

I had developed an interest in nuclear physics from these sources and when I started college I took courses in that field. I took a several year break for military service, where I became a radio operator.

After WW-II, I went back to college, but changed my specialization to electronics. I figured, with the "Bomb" developed, that nuclear physics would not be as interesting. I did take some courses in this area, such as nuclear instrumentation and atomic and molecular spectroscopy.

My working career was spent doing systems development work, mostly in electronics, but with a smattering of optics and mechanics. My early work was on classified projects that didn't require patent applications. In my later work developing consumer products, I received dozens of patents.

Since my retirement, I have had more time to think about really important things. I have gone back to my interest in cosmology, rereading my collection of books on relativity. Through the years of reading "Scientific American" I have watched the passage of the various theories on physics and astronomy.

Several years ago some articles triggered my desire to respond to what I considered a lack of logic in their premises. A couple of the articles made various uses of what Einstein considered to be his greatest mistake, the arbitrary coefficient that he introduced to make space not expand.

Another pair of articles appeared on adjacent pages of an issue. One article considered light as an electromagnetic wave; the other one treated light as photons imparting momentum. It struck me that there ought to be a way of combining these concepts. I will describe my thought on this later.

Now it is about time for me to start disclosing my thoughts about what I think the universe is all about. I will start at the present time, going from the smallest objects to the largest and then go back to the beginning, and then forward to the ending and back to the beginning.

Worlds Within Worlds

By
Lenard Metzger
Circa 1945

RELATIVITY
(Haiku for Albert)

We don`t see black holes
Because of their gravity
Light is too heavy

Our cosmic space curves
Warped by all the galaxies
Nothing can escape

Someone very far
Beyond our black universe
Won`t see us in here

By
Lenard Metzger
Circa 1985

Contents   Page

Introduction   163
Starlight
Stars
Black Holes
Big Bang
Multiverse
Gravity
Time   167
Life   173
Crackpot Idea   175
Conclusions   179
Addenda
  A. Light
  B. Radio
  C. Hydrogen Atoms
  D. Other Atoms
  E. Tables

INTRODUCTION

What do I mean by common sense cosmology? I believe that common sense should be used as the criteria in deciding between possible explanations about anything. I consider cosmology to be the study of everything, including our universe and its place in the cosmos. We will consider our universe, as it appears to exist at this time, and then look at where it came from and where it is going. To be able to do this we will have to establish some fundamental assumptions.

I believe that it is safe to assume that our universe began with what is called the "Big Bang". An expanding universe was postulated based on the red shift of various stars and galaxies, as a function of their distance from the Earth. Hubble and others quantified this relationship. The age of the universe was estimated at between 10 and 20 billion years. This was arrived at by projecting the estimated rate of expansion (the Hubble Coefficient) backward in time to where all the observed bodies would have coincided.

The measurement of the microwave background radiation of space and its equivalent black body temperature tended to verify the time required in cooling the initial, extremely high temperature to that which is measured now.

According to Einstein's theory, if the expanding universe had above a certain average density of matter within it, the universe would be what is deemed "closed".

The universe would eventually stop expanding and start to collapse back toward a common center. If the average density were appreciably less, the universe would be considered "open" and would keep expanding forever.

I believe that the latest estimate of the total visible and probable dark matter in the universe is less than the required amount to close the universe. It was hoped that the universe would be closed and that eventually all the matter would collapse back to the center and perhaps result in a new "Big Bang".

I believe it was suggested that this would take about a trillion years and that all of the stars would have died out. Most would have been absorbed into the central black holes of the galaxies. The remainder would be scattered individual black holes and neutron stars. Now, it is assumed that the expansion of the universe will continue without end.

Considerable theoretical work has been done on the earliest time of the universe. The present theory is that the explosion's temperature was so high initially, that only gravitational energy existed.

As the expansion continued and the temperature dropped, successive energy régimes occurred; strong force, weak force and electro-magnetic force. Soon, subatomic particles appeared, followed by the first atoms, hydrogen, deuterium and helium. Some postulate that a period of greatly increased rate of expansion occurred earlier.

As time went on the rapidly expanding sphere of hot gas changed from a homogeneous sphere into one with spaces and voids.

Over vast periods of time, the hydrogen gas converged, becoming numerous stars that combined into galaxies. Studies of the distribution of galaxies as a function of distance from us indicate that space is essentially flat.

The above concepts will be the starting point for my new ideas. The approach that will be taken is to use the least amount of mathematics possible and use simple logic to describe my vision of how the cosmos might work. Generally accepted astronomical and physical facts will be assumed as givens. Conclusions that have been reached as the result of manipulations of derived equations will be questioned and subject to re-interpretation. There are a number of prevalent theories that are not so easily understood or visualized. In some cases they are not even logical. In the following sections I will address these cases in detail and offer alternative explanations.

LIGHT DELIGHT

Bright white sunlight
Strikes the Earth's atmosphere
Ultra violet light
Triggers aurora borealis
Sky blue light
Refracts from the ozone layer
Sea green light
Filters down to sunken ships
Rusty red light
Reflects off of Mars
Razor sharp laser light
Bounces off the Moon
Billion-year-old quasar light
Red shifts with age
Invisible light
Stays inside a black hole

Lenard Metzger
Circa 1986

TIME

I have saved the "best" for last, TIME. Since I believe that time had no beginning and will have no end, the question is where to begin?

We all know what time is when we experience it in a straightforward way. The possibility of going back in time has been written about in fiction. Some scientists even suggested that it might be possible to do so. They speak of theoretical "worm-holes" that a space ship could enter and come out at a vastly different place and time. They have also described how particles created by collisions of atoms could be considered as moving backward in time and recombining into the original atoms. I do not believe that any of this is reasonable.

I believe that time is a linear dimension on which our mind places events in logical sequence. We can project backwards and forwards along the time line, using our imagination. This is subjective time. We calibrate our perception of time by how long it takes for our planet to revolve once a day, and to circle the Sun once a year. But there is a real objective time, in which past events occurred in sequence, without a mind being aware of their occurrence.

It has been suggested that time began with the start of our universe. I do not believe that this is true. My number one assumption is the following:
The passage of time had no beginning and will have no end. Time is infinite.

This takes us to a consideration of infinity. In elementary plane geometry we were given the property of two parallel lines that will extend to infinity and still not cross. An alternate form of geometry proposed that parallel lines did cross at infinity. This created the impression that infinity was a far distant place that could be reached. But this is nonsense. Things that extend infinitely never reach an end point. If things have been occurring for an infinitely long time, they never had a beginning.

The theory of relativity says that time varies with the velocity of an object. This idea first arose because of the belief in the "luminous ether." For a hundred years it was assumed that light, as a wave, needed a medium in which to travel through space. This medium was called the "luminous ether."

A problem arose in trying to measure the velocity of light. It had been assumed that the velocity of earth traveling through the ether would produce a change in the speed of the light moving in the direction of the motion of the earth. This was as compared to the speed measured at right angles to this direction.

Many attempts were made to detect this change in velocity, using instruments with ever increasing accuracy. No change in the velocity could be measured.

In a last effort to retain the concept of ether as the medium for electro-magnetic radiation, a theory was proposed that the units of length and time, used to measure the velocity of light, must change as a function of the velocity of light. The equations that described this effect are called Lorentz transformations.

Einstein used these equations in his theory of relativity. He applied them to all matter, except for light. He defined the velocity of light as constant. His equations have time slowing with increased velocity. He also derived equations for the relationship of mass to velocity and the equivalence of mass and energy.

The assumed time dependence on velocity has lead to many science fiction stories. The idea of a person traveling at near light speed and aging much more slowly than someone remaining behind has been intriguing. I believe that this cannot happen, for a number of reasons. We know of no practical way to accelerate a living person to a velocity where a significant change in mass and time would be predicted. Even if one could reach such a velocity, the physical limitations of a human body would not tolerate such changes. Muscles could not cope with a ten-ton body.

An apparent increase in mass with an increase in velocity could be explained by the effect of graviton interactions. Such an effect has been seen using particle accelerators. However, any apparent change in time, as measured with mechanical or even electronic clocks, could be due to the effects of gravitons on the clock mechanisms or atomic particles used for measuring time.

I believe that time is something sensed by the human mind. It is a subjective sensation that can be felt as fast or slow without reference to a clock. In physics, time should only be considered as the ratio of a distance relative to the velocity used in traversing the distance.

I don't believe that time should be treated as an independent variable, as it is in the theory of relativity.

The above is the crux of my difference with the theory of relativity. Relativity refers things to the point of view of observers. I believe that reality does not require observers for actual events and relationships to be true. I believe there is such a thing as absolute velocity in relation to a point in space. I am sure there are simultaneous occurrences throughout the universe, even though no observer can verify them.

BEFORE THE BEGINNING OF TIME

Lying awake some dark night
Try to imagine what it was like
When all of space was empty and black
Before the beginning of time

It takes a while to reach the state
Where darkness is an endless void
Stretching forever in every direction
Before the beginning of time

But feel the pulsing in your chest
Rising and falling with every breath
Telling time with a metronome heart
Before the beginning of time

Imagine being a bodiless mind
Conscious of only empty space
Still you will sense something pass
Before the beginning of time

Thinking one thought after another
Is somewhat like a ticking clock
You cannot think of timeless time
Before the beginning of time

Lenard Metzger
Circa 1986

LIFE

At the end of these writings I realized that I had omitted a very important item, life. No matter how wonderful the cosmos proves to be, if there were no one here to appreciate its beauty, it would be an empty process. Religion has an explanation for the start of life that satisfies many. Those who question that answer are still doubtful about some of the alternatives. It hardly seems possible that the complex form of life found on our world has had enough time to develop since the Earth became habitable. One suggestion has been made that this world was seeded from space. Then the question becomes, where did these seeds come from and how were they made?

It occurs to me that if beings were living at a time when their universe was dying, they might develop such seeds and disperse them into space. They could hope that their universe would become part of another universe and that some of the seeds would survive the Big Bang. If beings in some other dying universe did the same, also contributing seeds to the new universe, then many kinds of life could arise, with the possibility of something like cross-pollination.

We can assume that such advanced races will have perfected genetic engineering. Their seeds would probably incorporate many alternative genomes, programmed to allow for a vast range of environmental conditions.

The most likely vehicles for transporting such seeds would be on small bodies of ice, to provide the water necessary for the seeds to develop. They would also include the initial nutriments. These objects would be released in vast quantities.

Based on the animal population of earth, these progenitor races probably were mammalian, reptilian and insect-like types.

CRACKPOT IDEA

The following is the only idea that I am presenting that may truly be a crackpot idea. I am assuming that gravity is really caused by gravitons imparting momentum to atoms.

In that case, what would be nice to develop is a composite material or device that acts like an electronic rectifier or diode. A rectifier allows current to flow through it in one direction but not in the other. With such a device applied to gravitons, one could actually get an anti-gravity effect.

From one direction, most of the gravitons would pass through without imparting much momentum. From the other direction, they would impart momentum. This would move the device in the direction of the applied momentum.

In the following diagram (Figure 40) I have suggested how these devices could be used to propel a space ship. I have assumed that they will be in the form of flat panels that can be moved over a range of orientations independently. If the reflective side of the panels were facing toward the rear of the ship, they will be propelled forward. Four of these devices are shown as the dark objects in the bottom view.

After the ship has accelerated to the desired velocity, the devices can be rotated to the axial position, as shown by the one dark device in the side view. In the neutral case, every other panel will have a reversed orientation.

When it is desired to slow the ship down, devices can be rotated so that the reflecting surfaces are facing forward. To stay at any velocity, every other panel can face forward and the others can face the rear.

Another interesting feature would be that, when traveling at the desired velocity, the devices can be moved to the axial position, all facing in the same radial orientation. In this case they will respond to the gravitons coming from all the radial directions and cause the ship to spin around its fore and aft axis. This will result in centrifugal force providing weight to the crew in the area around the circumference of the ship. To stop adding spin, the panels can be alternated in their orientation so that every other one will respond to opposite momentum.

I have shown a central control area with a forward view. It can be arranged to have this central area free to counter-rotate relative to the circumference if it is desired. Many different combinations of the orientation of the panels are available to provide all the necessary control functions for the ship motion.

One last thought; if the ship were allowed to accelerate at a comfortable one g (the force of Earth's gravity) for about a year, it might reach a speed limited by the velocity of the gravitons. Wow, that would be speed of light travel!

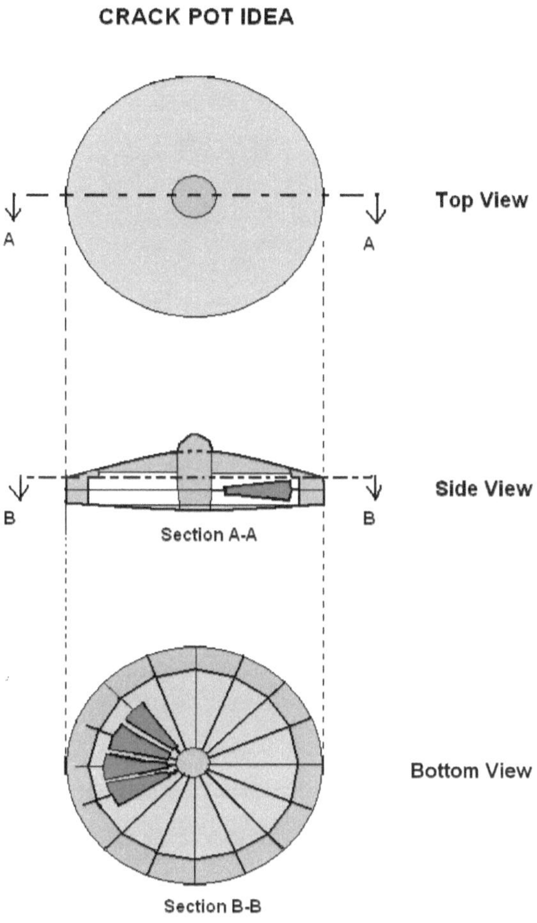

CRACK POT IDEA

Top View

A   A

Side View

B   B

Section A-A

Bottom View

Section B-B

Figure 40

CONCLUSIONS

With all due respect to Albert Einstein, I have presented my ideas of a worldview that may not be as mathematically elegant as his, but hopefully they will be more readily visualized. I see things differently then those who use his equations unquestionably as the starting point for their extrapolations. My ideas may not all prove to be correct, but they have the beauty of simplicity and being somewhat understandable.

Einstein was able to develop his Special Relativity in less than a year. It took him ten years to put together his General Relativity, to tie gravity in with his first approach. He did this with some borrowing from others. He then spent the rest of his life trying, unsuccessfully, to truly unify gravity with the other forces of nature.

In summarizing the ideas I presented here: I believe that the three dimensional view of objects moving out into a universe of non-expanding space is useful to overcome the dead end of the expanding space model. This model predicts that every other body in space will eventually disappear from our view, moving away at a speed faster than light.

I am very sure that our universe is a naturally occurring process that must have a means of reproducing itself. That is why I feel that the idea of multiple universes surrounding ours is reasonable.

The idea that photons have mass is not really too new. It is only the idea that they are limited to a velocity determined by gravitons that is new.

I must say that I consider the idea of gravitons, acting as a repellant force that imparts momentum to objects and produces the action of gravity, is the most significant of my ideas presented here. This also may be the most controversial.

I believe the concept of black holes being knowable as objects that are produced by a natural extension of known processes is a reasonable one. This ties together a progression; from a large star mass black hole, to a galaxy mass black hole, to an universe mass black hole, and to the Big Bang. It seems that the graviton idea is compatible with black holes, except for the mythical effect on time at the so-called event horizon.

I am the most uncomfortable about time being included in equations for explaining the effects of velocity on an object, such as changing its size and mass.

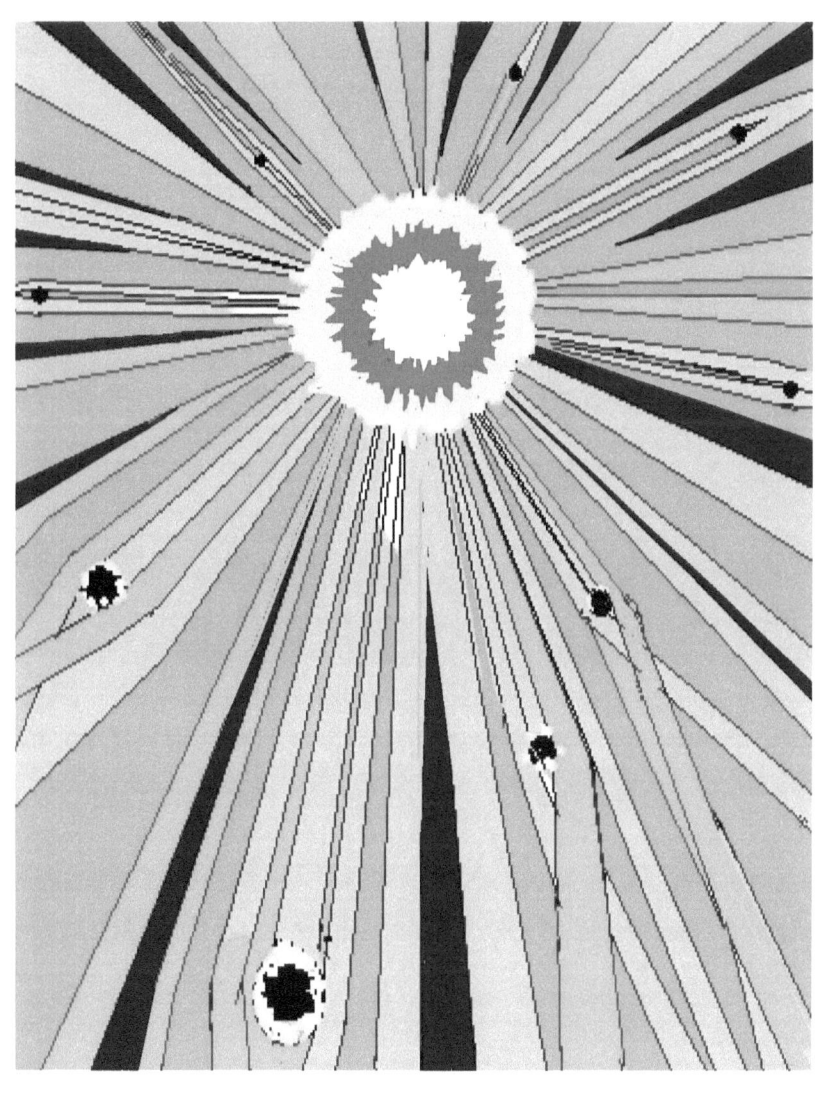

Big Bang and Black Hole Interactions

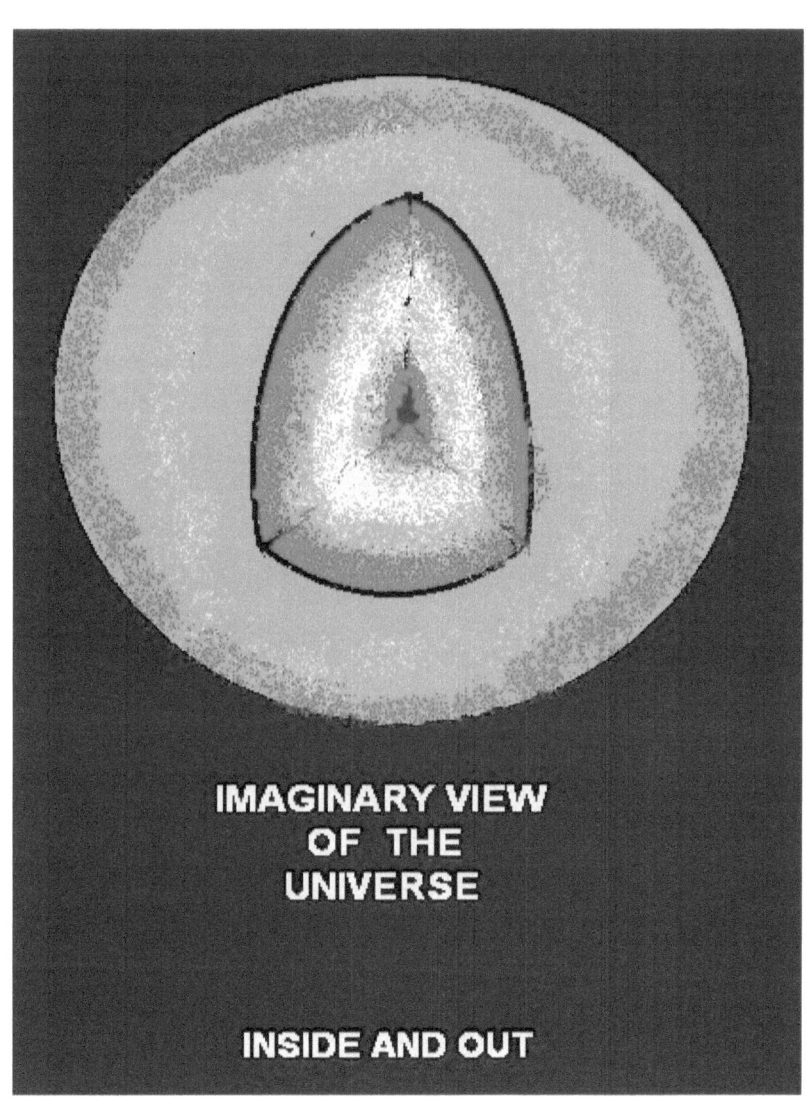

IMAGINARY VIEW
OF THE
UNIVERSE

INSIDE AND OUT

www.ingramcontent.com/pod-product-compliance
Lightning Source LLC
Chambersburg PA
CBHW030944180526
45163CB00002B/700